# Bioprospecting of Microbial Resources for Agriculture, Environment and Bio-chemical Industry

Pinki Saini • Pragya Mishra

# Bioprospecting of Microbial Resources for Agriculture, Environment and Bio-chemical Industry

Springer

Pinki Saini
Centre of Food Technology
Institute of Professional Studies
University of Allahabad
Prayagraj, Uttar Pradesh, India

Pragya Mishra
Food Processing and Management
DDU Kaushal Kendra, RGSC
Banaras Hindu University
Prayagraj, Uttar Pradesh, India

ISBN 978-3-031-63843-5 ISBN 978-3-031-63844-2 (eBook)
https://doi.org/10.1007/978-3-031-63844-2



This Springer imprint is published by the registered company Springer Nature Switzerland AG
The registered company address is: Gewerbestrasse 11, 6330 Cham, Switzerland

# Preface

Bioprospecting has never been more relevant and has gained renewed interest, because of the extreme rise in novel resistant pathogenic microorganisms. The systematic identification, assessment, and exploitation of the variety of life in a particular location is known as bioprospecting, and its goal is to find genetic resources for use in industry. In terms of economics, food security, sustainability, a temperate climate, efficient nutrient cycling, and cultural practices, the field can also greatly benefit ecosystems. Numerous distinct habitats are home to special microbes that may provide new antibiotics, enzymes, food, and energy sources. Even with all of the progress made in using microorganisms for human benefit, there is still much to learn about the diversity and interdependencies of microorganisms. The book emphasizes the limitations and opportunities related to the generation of microbial products. Various traditional and advanced bioprospecting technologies have been developed and employed to increase the knowledge on the microbiological diversity of functional processes in microbial ecosystems, and also for the identification and use of new biotechnological products. The current trend of creating new foods with improved health attributes is probably here to stay. A strategic study is required to ascertain the connection between processing and the decrease of harmful substances as well as to comprehend the dynamics of various processing techniques and how they might be adjusted to boost nutritional content while extending shelf life. To improve their production methods, conventional processing systems must use contemporary biotechnology approaches.

Numerous facets of bioprospecting microorganisms for use in pharmaceutical, nutraceutical, biomedical, and bioenergetic applications are covered in the book. The book provides a thorough understanding of the use of microorganisms or bioactive compounds derived from microorganisms for the creation of sustainable methods. It addresses current developments and difficulties in the synthesis and extraction of bioactive compounds, including enzymes, biopesticides, biofertilizers, biosensors, medicines, nutraceuticals, and pharmaceuticals. The difficulties posed by the various methods of microbial bioprospecting are discussed, as well as potential ways to get beyond these restrictions. There are also discussion on the use of microbe-based products in the development of greener technologies and the management of environmental pollutants. The book covers established knowledge as well as genomic-based technological advancements to offer a better understanding of its application to various industries. It also promotes the commercialization of microbially derived bioactive compounds and their application in biochemical industries, agriculture, and environmental protection studies. This will give readers valuable insight into the fundamentals of microbial prospecting.

As a whole the book contains 12 chapters. Both the authors have sincerely worked to explore the wider application in the major areas, including crop production, agriculture, health care, nonfood uses of crops and environmental issues including the areas of agriculture, the food business, and public health during the creation of the book's chapters. This book is appropriate for administrators, policy makers, consultants, industry persons, academicians, scientists, researchers, and graduate and post-graduate students working and studying in the areas of microbiology, food biotechnology, industrial microbiology, plant biotechnology, microbial biotechnology, environmental sciences, microbial prospecting, and biochemical engineering. The book will serve as an indispensable guide for anyone looking for a comprehensive overview of the subject.

Prayagraj, Uttar Pradesh, India

Pinki Saini
Pragya Mishra

# Acknowledgment

This reference book stands as a culmination of extensive research, collaboration, and dedication. We wish to express our sincere gratitude to those whose support has been instrumental in the realization of this project.

We are indebted to the numerous scholars and researchers whose groundbreaking work in the field has paved the way for the knowledge presented here. Your contributions have been a constant source of inspiration.

*We express our sincere gratitude to our teachers who have played an indispensable role in shaping our academic journey and personal growth.*

To our families, whose unwavering support and understanding have sustained us throughout the writing process, we express our deepest gratitude. Your encouragement has been our anchor.

Lastly, we want to thank the publishers and editorial team for their expertise and commitment to excellence. Your dedication to the dissemination of knowledge has made this book possible.

# Contents

# About the Authors

**Dr. Pinki Saini** is currently working as Associate Professor at Centre of Food Technology, University of Allahabad. She completed her M.Sc. degree in Food Science and Technology from G. B. Pant University of Agriculture and Technology, Pantnagar. She completed her Ph.D. in 2008 on "Preparation of Shatavari (*Asparagaus racemosus*) powder and its utilization as a preservative and therapeutic agent in Burfi" from G. B. Pant University of Agriculture and Technology, Pantnagar. She has qualified Agricultural Scientists Research Board (ASRB) National Eligibility Test Examination in Food Technology. She has 15 years of teaching and 18 years of research experience. She has supervised 6 Ph.D. theses and 45 M.Sc. dissertations. She has published more than 75 research/review papers in peer-reviewed journals and conferences. She has completed two sponsored research projects funded by Uttar Pradesh Council of Agricultural Research and granted one patent (Patent no.: 402984) by the Indian Patent Office. She has received an accession number MCC 4716 and deposited novel isolated strain as *Lactiplantibacillus plantarum* sp. strain ANPSCRLB at NCMR (Pune, Maharashtra). She has edited books on *Food and Industrial Bioprocessing*, published by New India Publishing Agency (NIPA), India, and *Biosorption Processes for Heavy Removal*, published by IGI Global, USA. She also has more than 18 book chapters to her credit in national and international published books. She is a life member of Association of Food Scientists and Technologists, India (AFSTI). Her research interests include bioprocessing of foods, fermentations studies, probiotics, biosorption of heavy metals, biofilm formation, and sustainable edible films from plant and microbial sources.

**Dr. Pragya Mishra** is currently working as Assistant Professor, Food Processing and Management Course, at DDU KAUSHAL Kendra, Banaras Hindu University. She completed her Master's degree in Food Technology in 2010 from the Centre of Food Technology, University of Allahabad with distinction. She did her dissertation research on; "Feasibility studies for enhancement of sensory attributes of pearl millet weaning food by fermentation", from the National Dairy Research Institute, Karnal, Haryana. She has qualified University Grant Commission National Eligibility Test Examination and awarded by Doctor of Philosophy in Food Technology from Centre of Food Technology, Institute of Professional Studies, University of Allahabad, India, in 2017, titled "Isolation and molecular characterization of edible Spirulina species under different abiotic stress condition and evaluation of their food supplement." She worked as Senior Research Fellow of Uttar Pradesh Council of Agricultural Research funded project, entitled "Evaluation and promotion of processing varieties of potato in selected regions of U.P. and development of potato-based RTE products." She had received an accession number (GenBank: KM407008.1) from NCBI, USA, for new isolates of *Spirulina maxima* species from cyanobacterial isolates targeting specific 16sRNA gene. She had been granted a patent on "Potato flakes energy bar" (Patent no.: 402984) by the Indian Patent Office. She has published more than 40 international and national research/review articles in peer-reviewed journals; 1 edited book published by Springer Nature, Singapore; 5 edited books published by Bharti Publication, New Delhi; and contributed 12 chapters in international books. She had received MFPA Technology, 2018, Young Scientist award, organized by Banaras Hindu University. She also had received Best Female Researcher Award-2022 on Women's day 2022 organized by RGSC, Banaras Hindu University. She is Life Time Member of All India Medical Laboratory Technologists Association, New Delhi, and Society of Bacteriophage, Varanasi

# Chapter 1
# Microbial Bioprospecting for Nutraceuticals as Novel Therapeutics

## Contents



## 1.1 Introduction

Nutraceuticals are a group of bioactive compounds with diverse structural and functional properties that offer long-term medicinal or physiological benefits beyond their nutritional or direct pharmaceutical effects. They are different from functional foods and drugs and can be derived from a variety of sources such as plants, animals, microorganisms, and marine sources. Nutraceuticals are widely used for their




P. Saini, P. Mishra, *Bioprospecting of Microbial Resources for Agriculture, Environment and Bio-chemical Industry*, https://doi.org/10.1007/978-3-031-63844-2_1

health-promoting or disease-preventing properties, particularly in preventing aging-related diseases such as oxidative stress, depression, inflammation, arthritis, osteoporosis, gastrointestinal diseases, cardiovascular diseases, diabetes, and cancer. They can be derived from plants (e.g. phytochemicals, vitamins), from animals (e.g. polysaccharides), from microorganisms (e.g. poly amino acids) and from marine sources (e.g. glucosamine and chitosan). Plant-based nutraceuticals are typically only produced in low quantities during times of infection or stress, and their complex structures and stereochemical features make chemical synthesis challenging. As a result, it is difficult to produce these bioactive compounds in large quantities. Although researchers have attempted to engineer the metabolic pathways in plants to increase the production of desired compounds, modifying plant metabolism is a complex and difficult process, as noted by Madhavan et al. (2019). However, the extraction of nutraceuticals from natural sources can be challenging and is associated with low product yield. This has led to the development of microbial production platforms as a sustainable and eco-friendly alternative for the large-scale production of value-added nutraceuticals from simple carbon sources. Microbial platforms such as *Escherichia coli* and *Saccharomyces cerevisiae* have been engineered to produce a wide range of complex and diverse value-added chemicals, including phytochemicals, prebiotics, polysaccharides, and polyamino acids. Synthesizing natural bioactive compounds in microbes using different strategies has been successful. The demand for nutraceuticals has been on the rise due to the increasing interest in maintaining human wellness through diet. The nutraceutical market is estimated to be worth over $230 billion globally, with a significant portion of it being in the US, where it was valued at $75.9 billion in 2019 (Jain & Ramawat, 2013).

## 1.2 Need for Microbial Production of Nutraceuticals

The conventional nutraceutical industries cannot keep up with the expanding nutraceutical market due to limited availability and high cost of raw materials, low content and purity of nutraceuticals, and complicated extraction procedures. Chemical synthesis is an alternative approach but is limited to producing simple biochemicals and is not feasible for complex biochemicals, especially those that are chemically unfavorable (De Luca et al., 2012). Moreover, the yield of plant-based nutraceuticals is low due to their production only during infections and stress conditions, and their stereochemical features and composite structures make chemical synthesis difficult. Although metabolic engineering of plant metabolism has been attempted, it is complex and stringent (Madhavan et al., 2019).

To address these issues, microbial-based metabolic engineering is a promising approach that has made significant progress in producing value-added nutraceuticals. Microbial production of nutraceuticals is considered eco-friendly and sustainable. The elucidation of biosynthetic pathways and genetic manipulability of microorganisms have enabled the development of microbial hosts for production of various nutraceuticals. Genetic manipulation and optimization of culture conditions or fermentation processes make it possible to scale up the production of nutraceuticals from simple carbon sources. Commercial microbial food-grade strains with Generally Recognized As Safe (GRAS) status have been used as cell factories to produce natural products. Metabolic modification of microbial hosts for nutraceutical compounds offers an alternative to chemical routes and extraction, allowing clean products to be synthesized at mild conditions without high heat and pressure. This approach is more efficient and sustainable than direct extraction or chemical synthesis and has great potential for producing high-quality nutraceuticals to meet the growing demand (Luo et al., 2015).

## 1.3 Phytochemicals

Phytochemicals are bioactive secondary metabolites derived from plants, found in a variety of fruits, vegetables, beans and grains. They play an important role in plant defense against biotic and abiotic stresses, and have been shown to have health benefits for humans. Nutraceuticals benefit greatly from phytochemicals, which include polyphenolic compounds such as flavonoids, isoflavonoids, stilbenoids and curcuminoids, alkaloids, terpenoids such as monoterpenes, diterpenes, tetraterpenes, polyterpenes, steroid saponins, lycopene and carotenoids, as well as their derived compounds. These compounds are typically derived from the phenylpropanoid pathway. Natural polyphenols are often present in conjugated forms with other polyphenols, organic acids, or carbohydrates, making them difficult and inefficient to isolate and purify using conventional extraction methods. In microbial production of phytochemicals, pathway engineering plays a crucial role. Through genetic engineering techniques, scientists can introduce or modify specific metabolic pathways within the microorganism to enable the production of desired phytochemicals. This involves identifying and incorporating genes encoding enzymes involved in the synthesis of the target compound. Table 1.1 illustrates the different nutraceuticals produced through their microbial host, utilized substrate and production yield.

### *1.3.1 Polyphenolic Compounds*

Polyphenolic compounds are a class of bioactive secondary metabolites that are commonly found in plants. They exhibit a wide range of biological activities, including antioxidant, anti-inflammatory, anti-cancer, anti-diabetic, anti-obesity, anti-neurodegenerative, and anti-microbial effects. These compounds can be found in various plant sources such as fruits, vegetables, beans, grains, and herbs. The structural diversity of polyphenols is based on their fundamental phenolic structure, which can be altered through processes such as hydroxylation, methylation, glycosylation, and acylation (De Luca et al., 2012).

**Table 1.1** Microbial production of nutraceuticals

| Nutraceuticals | Microbial host | Substrate | Yield (mg/L) | References |
|---|---|---|---|---|
| **Polyphenolics** | | | | |
| Apigenin | *E. coli* | Tyrosine | 13 | Miyahisa et al. (2006) |
| Galangin | *E. coli* | Tyrosine | 1.1 | |
| Chrysin | *E. coli* | Phenylalanine | 9.4 | |
| Kaempferol | *E. coli* | Phenylalanine | 15.1 | |
| Genistein | *E. coli-S. cerevisiae* | Tyrosine | 6.0 | Katsuyama et al. (2007) |
| Catechin | *E. coli* | Eriodictyol | 910.9 | Zhao et al. (2015) |
| Afzelechin | *E. coli* | Glycerol | 41 | Jones et al. (2016) |
| Curcumin | *E. coli* | Tyrosine | 0.6 | Wang et al. (2015) |
| Pinocembrin | *E. coli* | Glucose | 40.2 | Wu et al. (2013a, b) |
| Eriodictyol | *E. coli* | Tyrosine | 107 | Zhu et al. (2014) |
| Resveratrol | *E. coli* BW27784 | p-Coumaric acid | 0.16 | Afonso et al. (2014) |
| | *E. coli* BL21(DE3) | p-Coumaric acid | 16 | Beekwilder et al. (2006) |
| | *E. coli* BRB | Cinnamic acid | 155 | Katsuyama et al. (2007) |
| | *E. coli* BL21(DE3) | Glucose | 304.5 | Wu and Shah (2017) |
| | *C. glutanicum* | p-coumaric acid + cerulenin | 158 | Kallscheuer et al. (2016) |
| | *C. glutanicum* | Glucose + cerulenin | 59 | Braga et al. (2018) |
| | *S. cerevisae* | p-Coumaric acid | 3.1 | Shin et al. (2011) |
| | *S. cerevisae* | p-Coumaric acid | 5.25 | Zhang et al. (2006) |
| Naringenin | *E. coli* BL21 (DE3) | Glucose | 100.64 | Wu et al. (2014) |
| | *S. cerevisiae* | Glucose | 54.4/112.9 | Koopman et al. (2012) |
| | *E. coli* BW25113 and *S. cerevisiae* CEN.PK2-1C | Xylose | 21 | Zhang et al. (2017) |
| Quercetin | *S. cerevisiae* CEN.PK102–5B | Glucose | 20 | Rodriguez et al. (2017) |

(continued)

**Table 1.1** (continued)

| Nutraceuticals | Microbial host | Substrate | Yield (mg/L) | References |
|---|---|---|---|---|
| **Carotenoids** | | | | |
| Astaxanthin | *E. coli* | Glucose | 1.4 mg/g | Lemuth et al. (2011) |
| | *S. cerevisiae* BY4742 YPD | Glucose | 217.9 | Jin et al. (2018) |
| | *Y. lipolytica* GB20 | YPD medium | 54.6 | Kildegaard et al. (2017) |
| Zeaxanthin | *E. coli* | Glucose | 43.46 | Li et al. (2015) |
| β-Carotene | *E. coli* ATCC 8739 | Glycerol | 2100 | Zhao et al. (2013) |
| | *E. coli* BL21(DE3) | Glycerol | 3200 | Yang and Guo (2014) |
| | *Y. lipolytica* MYA2613 | Glucose | 4000 | Gao et al. (2017) |
| | *Y. lipolytica* Po1d | YPD medium | 6500 | Larroude et al. (2018) |
| Lycopene | *E. coli* | Glucose | 18 mg/g | Alper et al. (2005) |
| | *E. coli* BL21(DE3) | Glycerol | 1230 | Zhu et al. (2015) |
| | *S. cerevisiae* CEN.PK2-1D | Glucose or ethanol | 2370 | Ma et al. (2018) |
| **Alkaloids** | | | | |
| (S)-reticuline (Benzylisoquinoline alkaloid) | *E. coli* | Glycerol | 46 | Nakagawa et al. (2011) |
| Strictosidine (monoterpene indole alkaloid) | *S. cerevisiae* | Glucose | 0.53 | Brown et al. (2015) |
| Indolylgluco-sinolate (Glucosinolate) | *S. cerevisiae* | Glucose | 1.07 | Mikkelsen et al. (2012) |
| β-Alanine | *E. coli* W3110 | Glucose | 32,300 | Song et al. (2015) |
| GABA | *E. coli* BW25113 | M9 medium | 4800 | Soma et al. (2017) |
| | *C. glutamicum* ATCC 13032 | GP1 medium | 39,000 | Choi et al. (2015) |

(continued)

**Table 1.1** (continued)

| Nutraceuticals | Microbial host | Substrate | Yield (mg/L) | References |
|---|---|---|---|---|
| **PUFA** | | | | |
| EPA | *Y. lipolytica* ATCC 20362 | Glucose | 56.6% in total lipids with 15% DCWa | Xue et al. (2013) |
| | *Y. lipolytica* Y4305 | Glucose | 50% in total lipids with 25% DCWa | Xie et al. (2015) |
| | *E. coli* EPI300T1 | – | 14% | Amiri-Jami and Griffiths (2010) |
| | *E. coli* DH5α | – | 2.2 mg/g | Peng et al. (2016) |
| | *E. coli* DH5α | | 12% | Orikasa et al. (2009) |
| DHA | *E. coli* DH5α | LB medium | 2.4 | Peng et al. (2016) |
| | *Y. lipolytica* Y4305 | Glucose | 5.6% in total lipids | Damude et al. (2016) |
| | *Y. lipolytica* Y4305 | | 50% in total lipids with 25% DCW | Xie et al. (2015) |
| α-Linolenic acid | *Y. lipolytica* L36DGA1 | YSC medium | 1400 | Cordova and Alper (2018) |
| **Prebiotics** | | | | |
| 2-Fucosyllactose | *E. coli* | Lactose and glycerol | 20,280 | Baumgartner et al. (2013) |
| Galactooligosaccharides | *Lactococcus lactis* | Lactose | 197,000 | Yu and O′ Sullivan (2014) |
| **Polysaccharides** | | | | |
| Hyaluronan | *Streptomyces albulus* | Sucrose | 6200 | Yoshimura et al. (2015) |
| Chondroitin | *E. coli* | Glucose | 2400 | He et al. (2015) |
| Heparosan | *E. coli* | Glycerol | 1880 | Zhang et al. (2012) |
| Scleroglucan | *Sclerotium rolfsii* | Uridine mono-phosphate | 22,320 | Survase et al. (2007) |
| **Poly amino acids** | | | | |
| Poly-ε-L-lysine | Streptomyces sp.M-Z18 | Glucose and glycerol | 35,140 | Chen et al. (2012, b) |
| Poly-γ-glutamic acid | *Bacillus subtilis* BL53 | L-glutamic acid and glycerol | 17,000 | Da Silva et al. (2014) |
| Cyanophycin | *E. coli* | Yeast extract | 120 mg/g | Tseng et al. (2012a, b) |

Polyphenols are synthesized via the phenylpropanoid pathway, a metabolic pathway that produces a variety of secondary metabolites, including polyphenols. This pathway begins with the conversion of the amino acids L-phenylalanine and L-tyrosine into cinnamic acid and p-coumaric acid, respectively, by enzymes called phenylalanine ammonia lyase (PAL) and tyrosine ammonia lyase (TAL). These intermediates are then converted into cinnamoyl-CoA and p-coumaroyl-CoA by an enzyme called 4-coumaroyl-CoA ligase (4CL). Further conversion of these compounds is carried out by type III polyketide synthases, such as chalcone synthase (CHS), curcuminoid synthase (CUS), or stilbene synthase (STS). These enzymes transform the compounds into backbone flavanones, stilbenes, or curcuminoids (Lin et al., 2014). The final step in the biosynthesis of polyphenolic compounds involves the action of tailoring enzymes, including hydroxylases, O-methyltransferases, glucosyltransferases, lipases, and prenyltransferases. These enzymes modify the compounds to generate diverse polyphenolic compounds (Mora-Pale et al., 2013). *E. coli* and *S. cerevisiae* are commonly employed as host organisms for the production of polyphenolic compounds through de novo or semi-de novo approaches with advantages such as rapid growth and genetic manipulability in *E. coli*, while *S. cerevisiae* is considered Generally Regarded As Safe (GRAS) and capable of expressing plant metabolic enzymes. To facilitate the microbial production of polyphenolic compounds, enzymes derived from plants are often overexpressed using techniques such as codon optimization or the creation of enzyme chimeras for efficient bacterial expression (Leonard & Koffas, 2007).

In the first approach, heterologous hosts are supplemented with precursor molecules to enable the production of specific polyphenolic compounds. For instance, when phenylalanine or tyrosine is provided to *E. coli* strains expressing heterologous pathways, it can result in the production of compounds like chrysin (9.4 mg/L), galangin (1.1 mg/L), apigenin (13 mg/L), or kaempferol (15.1 mg/L) (Miyahisa et al., 2006). This strategy is particularly useful when precursor availability is limited or when these precursors cannot be synthesized within the microbial host. Additionally, by introducing glycosyltransferases with specific substrate preferences, flavonoid glucosides like luteolin 4′-O- and 7′-O-glucosides have been successfully produced in *E. coli* through supplementation with luteolin (He et al., 2008). Other studies have reported the production of quercetin derivatives, such as quercetin 3-O-glucoside, quercetin-3-O-glucuronide, quercetin 3-O-galactoside, and quercetin-3-O-rhamnoside, by expressing glycosyltransferases with different substrate specificities (Kim et al., 2015; Kim et al., 2012). Alternatively, the de novo biosynthesis of polyphenolic compounds from simple carbon sources like glucose and glycerol is an attractive approach for industrial applications. *E. coli* can be easily engineered to enhance the production of L-phenylalanine and L-tyrosine from these carbon sources. The intracellular concentration of malonyl-CoA, an extender unit for polyphenolic backbones, can be significantly increased in *E. coli* by overexpressing acetyl-CoA carboxylase (ACC), optimizing the metabolic network at the genome-scale, or repressing malonyl-CoA consumption pathways using antisense RNA (Xu et al., 2011; Yang et al., 2015). Successful de novo biosynthesis of certain polyphenolic compounds, such as pinocembrin, pinosylvin, sakuranetin, and ponciretin, has been achieved in *E. coli* (Wu et al., 2013a, b; Kim et al., 2013). Furthermore, the substrate promiscuity of native or engineered enzymes offers opportunities to produce novel or non-natural polyphenolic compounds (Bhan et al., 2015).

### *1.3.2 Alkaloids*

Alkaloids encompass a diverse group of natural compounds known for their various biological activities, such as anti-inflammatory, anti-tumor, and anti-microbial properties. While primarily found in plants, certain microorganisms are also capable of producing alkaloids. The biosynthesis of alkaloids in microorganisms often involves the shikimate pathway, a common metabolic route for the production of aromatic amino acids, including tryptophan. Several microorganisms, including *Escherichia coli, Streptomyces*, and *Pseudomonas putida*, have been utilized for the microbial production of alkaloids. For instance, the biosynthesis of camptothecin, an indole alkaloid with anticancer properties, has been achieved in *E. coli* by introducing a biosynthetic pathway derived from the plant *Camptotheca acuminata*. Similarly, other alkaloids like berberine, ajmalicine, and strictosidine have been successfully produced in microorganisms by introducing corresponding biosynthetic pathways (Nakagawa et al., 2011).

The initial step in alkaloid biosynthesis involves the condensation of two molecules of L-tyrosine to form L-dopa, catalyzed by tyrosine decarboxylase. L-dopa is then converted to dopamine by L-dopa decarboxylase. Dopamine is further transformed into the precursor of various alkaloids, 4-hydroxyphenylacetaldehyde, through the action of dopamine hydroxylase, followed by oxidation to 4-hydroxyphenylacetic acid. The next stage includes the condensation of 4-hydroxyphenylacetic acid with acetyl-CoA, catalyzed by 4-hydroxybenzoate synthase, to form 4-hydroxybenzoic acid. In some alkaloid biosynthetic pathways, 4-hydroxybenzoic acid is subsequently converted to the alkaloid backbone precursor, such as anthranilic acid, by anthranilate synthase (Nakagawa et al., 2014). The biosynthesis of alkaloids then involves intricate chemical modifications of the backbone, including oxidation, reduction, and methylation, catalyzed by a series of enzymes like cytochrome P450 monooxygenases, reductases, and methyltransferases.

There are two well-known types of alkaloids: benzylisoquinoline alkaloids (BIAs) derived from tyrosine, monoterpene indole alkaloids (MIAs) derived from tryptophan, and glucosinolates. Recent research on BIA biosynthetic pathways has enabled the reconstruction of these pathways in both *E. coli* and *S. cerevisiae*, resulting in the biosynthesis of (S)-reticuline from simple carbon sources with a titer of 46.0 mg/L. Additionally, bacterial consortia have been engineered to produce magnoflorine and scoulerine in *E. coli* and S. cerevisiae, respectively (Hawkins & Smolke, 2008). *S. cerevisiae* has also been modified to produce (R,S)-reticuline, scoulerine, tetrahydrocolumbamine, tetrahydroberberine, and salutaridine.

However, the metabolic engineering of MIA alkaloids in microbes is currently limited. Recent advancements involve the engineering of yeast to produce the MIA alkaloid strictosidine by introducing 21 new genes and deleting 3 genes in the yeast genome. In another rare example, *S. cerevisiae* has been used to produce tryptophan-derived indolylglucosinolate (IG) by integrating eight plant genes into its genome. Although plant platforms are better suited for scalable alkaloid production, metabolic engineering in microbes shows promising potential for cost-efficient synthesis of these plant-derived compounds (Brown et al., 2015).

### *1.3.3 Terpenoids*

Terpenoids represent a diverse class of natural compounds with a wide range of biological activities, including antitumor, antibacterial, and antifungal properties. While terpenoids are commonly found in plants, their extraction can be challenging and costly. However, advancements in metabolic engineering have enabled the production of terpenoids in microbes by manipulating their biosynthetic pathways. One strategy involves engineering the mevalonate pathway, which is a central pathway for isoprenoid biosynthesis in many microorganisms. This pathway converts acetyl-CoA into isopentenyl pyrophosphate (IPP) and dimethylallyl pyrophosphate (DMAPP), the fundamental building blocks for terpenoid production. By overexpressing key enzymes in this pathway, such as HMG CoA reductase, and combining them with heterologous terpene synthases, microbial strains can synthesize various terpenoids. For instance, microbial strains like *E. coli* and *S. cerevisiae* have been engineered to produce artemisinic acid, a sesquiterpene and precursor for the antimalarial drug artemisinin. This was accomplished by introducing a heterologous mevalonate pathway along with amorphadiene synthase and amorpha-4,11-diene monooxygenase, which converted artemisinic acid into artemisinin. Similarly, other terpenoids such as taxadiene, squalene, and geraniol have been successfully synthesized in different microbial hosts using similar approaches.

Terpenoids can be derived from either isopentenyl pyrophosphate (IPP) or dimethylallyl pyrophosphate (DMAPP) and encompass various subclasses, including monoterpenes (C10), sesquiterpenes (C15), diterpenes (C20), triterpenes (C30), tetraterpenes (C40), and polyterpenes (Marienhagen & Bott, 2013). Metabolic engineering has been employed to produce valuable terpenoid pharmaceuticals such as the anticancer drug paclitaxel and the antimalarial drug artemisinin. In the nutraceutical industry, tetraterpene carotenoids like β-carotene, lutein, lycopene, α-carotene, and astaxanthin find applications as natural food colorants and feed supplements. Heterologous hosts such as *E. coli* and *S. cerevisiae* have been employed for combinatorial carotenoid biosynthesis, and extensive efforts have been made to optimize production strains. For instance, lycopene production in *E. coli* was enhanced through systematic and combinatorial gene knockouts, as well as engineering central metabolic modules to increase ATP and NADPH supplies, resulting in β-carotene production with a yield of 60 mg/g DCW (dry cell weight) (Lin et al., 2014).

Zeaxanthin production was achieved through biosynthesis utilizing lycopene, and its titer was improved to 43.46 mg/L by coordinating the expression of the lycopene β-cyclase gene crtY and the β-carotene 3-hydroxylase gene crtZ via protein fusion approaches. Furthermore, a plasmid-free *E. coli* strain with integrated xanthophyll biosynthetic genes successfully produced astaxanthin with a yield of 1.4 mg/g DCW (Lemuth et al., 2011). Novel carotenoids, including 4-ketozeinoxanthin and rare C50 carotenoids like sarcinaxanthin, decaprenoxanthin, and sarprenoxanthin, have also been synthesized in lycopene-producing *E. coli* hosts (Netzer et al., 2010; Maoka et al., 2014). Recently, a non-natural purple carotenoid C50-astaxanthin was produced in *E. coli* by introducing mutants of the first two pathway enzymes, farnesyl diphosphate synthase (FDS), and carotenoid synthase (CrtM), with modified substrate selectivities to enhance C50 backbone production and suppress non-target compounds (Furubayashi et al., 2015).

In a study conducted by Kim et al. (2017), a quorum sensing (QS)-based promoter system was successfully utilized for inducer-free production of bisabolene in *E. coli*. Through the optimization of LuxR and LuxI expressions, the titer of bisabolene reached 1.1 g/L. This approach demonstrates the potential of using QS-based systems for efficient terpenoid production. Streptomyces species have been a rich source of terpenoids, as highlighted by studies conducted by Koksal et al. (2012) and Rinkel et al. (2017). Therefore, Streptomyces bacteria hold promise as potential hosts for terpenoid production in the future (Phelan et al., 2015). Synthetic biologists have also designed promoters that respond to environmental signals such as pH and QS molecules (Tsao et al., 2010; Zargar et al., 2015; Rajkumar et al., 2016). These environmentally responsive promoters offer a means to control terpenoid biosynthesis without relying on expensive inducer molecules. *Yarrowia lipolytica*, a yeast species, is another promising host for terpenoid production based on the mevalonate (MVA) pathway, thanks to its abundant acetyl-CoA, which is the initial substrate for the MVA pathway (Zhu & Jackson, 2015). Additionally, carotenogenic *Rhodosporidium toruloides* has been engineered to produce terpenoids from lignocellulosic hydrolysates. This yeast demonstrates the ability to utilize multiple carbon sources and exhibits tolerance to high osmotic stress, making it a valuable candidate for terpenoid production (Yaegashi et al., 2017; Sundstrom et al., 2018).

### *1.3.4 Resveratrol*

Resveratrol, a stilbene phytoalexin, is renowned for its various therapeutic effects, including antioxidant, anticancer, anti-inflammatory, and neuroprotective properties, as well as its positive impact on skin health, diabetes, and longevity. To meet the growing demand for resveratrol, considerable efforts have been directed towards its production using microorganisms as bioreactors. Utilizing microbial cells for resveratrol biosynthesis offers numerous advantages, such as rapid production, cost-effective cultivation media, ease of genetic manipulation, and the availability of protein and metabolic engineering tools. Over the past two decades, bacterial hosts like *Escherichia coli, Lactococcus lactis*, and *Corynebacterium glutamicum*, as well as yeast hosts such as *Saccharomyces cerevisiae* and *Yarrowia lipolytica*, have been engineered and employed for resveratrol production.

In addition to selecting suitable host organisms, several factors significantly impact resveratrol production, including the choice of pathway and non-pathway genes for transfer, codon optimization, pathway engineering, mutagenesis, protein engineering, genetic system used, vector designs, and synthetic scaffolds (Ibrahim & Yan, 2022). Careful consideration and implementation of these strategies play a crucial role in achieving high resveratrol titers. Resveratrol functions as a phytoestrogen receptor agonist, exhibiting the ability to suppress the expression of cyclooxygenase-2 (COX-2), an enzyme that promotes tumor growth. It has also been observed to play a role in preventing cardiovascular and neurodegenerative diseases (Nakata et al., 2012; Zhang et al., 2010). Biochemically, the production of resveratrol originates from the same precursor, p-coumaric acid, as described for naringenin. The conversion of coumaroyl-CoA into resveratrol is catalyzed by a stilbene synthase (STS), a type III polyketide synthase, using three molecules of malonyl-CoA.

Resveratrol can be efficiently synthesized from inexpensive carbon sources by employing the tyrosine pathway for p-coumaric acid production in *Saccharomyces cerevisiae*. This approach has yielded up to 531 mg/L of resveratrol in fed-batch fermentation (Li et al., 2015). To achieve this, the resveratrol biosynthetic genes (TAL from H. aurantiacus, 4CL1 from *A. thaliana*, and VST1 from *V. vinifera*) were integrated, and mutants with multiple chromosomal copies of ARO4K229L (feedback-inhibition resistant DAHP synthase), ARO7G141S (feedback-inhibition resistant chorismate mutase), and ACC1S659A, S1157A (inactivation-resistant acetyl-CoA carboxylase) were introduced (Li et al., 2015). The same research group also demonstrated de novo resveratrol synthesis from glucose or ethanol via the phenylalanine pathway in *S. cerevisiae* (Li et al., 2016). In fed-batch fermentation, these engineered yeast strains produced 812 and 755 mg/L of resveratrol from glucose and ethanol feed, respectively. Furthermore, *Streptomyces venezuelae* has been genetically modified to produce various bioactive natural compounds, including flavonoids and stilbenes. The manipulation of Streptomyces sp. was first reported in 2009 by Park et al., who expressed heterologous phenylpropanoid biosynthetic pathway genes (Park et al., 2009). They utilized 4-CL from *S. coelicolor* (ScCCL) and a codon-optimized STS from *Arachis hypogaea*. However, the engineered strain produced less than 0.4 mg/L of resveratrol.

In addition to yeast, an engineered strain of *Corynebacterium glutamicum* has gained popularity for the production of small molecules, including resveratrol (Zhang et al., 2019; Wang et al., 2019a, b). The engineered *C. glutamicum* strain (ΔphdB, ΔpcaF, and ΔpobA) was developed to produce resveratrol from p-coumaric acid and achieved a titer of 158 mg/L of resveratrol when the fatty-acid biosynthesis inhibitor cerulenin was added to the medium (Kallscheuer et al., 2016). More recently, Braga and colleagues designed a system to successfully produce resveratrol in a *C. glutamicum* DelAro4 strain using glucose while supplementing cerulenin, an inhibitor of fatty acid synthase, to increase the pool of malonyl-CoA within the host organism (Braga et al., 2018; Van Summeren & Marienhagen, 2015). Other organisms such as *Aspergillus niger, Lactobacillus lactis*, and *Aspergillus oryzae* have been utilized for the bioproduction of resveratrol by incorporating heterologous pathway genes including phenylalanine ammonia lyase (PAL), cinnamate-4-hydroxylase (C4H), 4CL from *Arabidopsis thaliana*, and STS from *Rheum tataricum* (Katz et al., 2015). Another approach involves an *E. coli* co-culture system, where two engineered *E. coli* strains are employed to produce resveratrol from glycerol. The first strain utilizes glycerol as a carbon source to synthesize p-coumaric acid, which is then excreted into the medium. The second engineered *E. coli* strain is rewired to efficiently produce malonyl-CoA and the STS enzyme, resulting in the production of 22.6 mg/L of resveratrol (Camacho-Zaragoza et al., 2016). These studies highlight the diverse range of microbial hosts and strategies employed for the biosynthesis of resveratrol, offering potential avenues for cost-effective and sustainable production of this valuable compound.

### *1.3.5 Naringenin*

Naringenin is a valuable flavonoid known for its various beneficial properties such as antioxidant, antibacterial, antitumor, cardioprotective, anti-inflammatory, and antiviral effects. However, extracting naringenin from plant sources or synthesizing it chemically can be complex and yield limited amounts. Consequently, microbial synthesis of naringenin is preferred over extraction and chemical methods (Pandey et al., 2016). The biosynthesis of naringenin begins with p-coumaroyl-CoA, which is derived from tyrosine and phenylalanine. This compound, along with three molecules of malonyl CoA, undergoes cyclization with the assistance of chalcone synthase to form naringenin chalcone. Subsequently, chalcone isomerase catalyzes the conversion of naringenin chalcone into naringenin. Therefore, efforts in strain engineering have focused on optimizing these steps.

To enhance the supply of p-coumaric acid within the cell, two different biosynthetic pathways can be utilized. In one pathway, phenylalanine ammonia lyase (PAL) converts phenylalanine into cinnamic acid, which can then be hydroxylated into p-coumaric acid by cytochrome P-450-dependent cinnamate-4-hydroxylase (C4H) (Pandey et al., 2016). However, expressing heterologous P450 enzymes and finding compatible reductase partners can be challenging, particularly in bacterial cells (Hwang et al., 2003). An alternative route involves the direct conversion of tyrosine into p-coumaric acid through the action of tyrosine ammonia lyase (TAL) (Rodriguez et al., 2017). Regardless of the chosen pathway, the resulting p-coumaric acid is then converted into its Coenzyme A ester, coumaroyl-CoA, by 4-coumarate:CoA ligase (4CL) (Palmer & Alper, 2018).

*E. coli* strains have been engineered to enhance naringenin production by overexpressing phenylalanine ammonia lyase (PAL), 4-coumarate CoA ligase (4CL), and chalcone synthase (CHS) genes from *Rhodotorula rubra, Streptomyces coelicolor*, and *Glycyrrhiza echinate*, respectively (Hwang et al., 2003). Another study by Zhou et al. (2020) involved the overexpression of aroGfbr and tyrAfbr genes, knockout of the tyrR repressor gene, and optimization of glucose, glycerol, and potassium acetate levels in the medium, resulting in naringenin production of up to 588 mg/L in *E. coli*. Ganesan et al. (2017) developed a co-culture system with *E. coli*, leading to the production of 41.5 mg/L of naringenin. Parra Daza et al. (2023) employed parameterized algorithms to identify the shortest pathway for (2S)-naringenin production and selected the genes pal, 4 cl, chs, and chalcone isomerase (chi) for the biosynthetic pathway. They achieved a yield of up to 588 mg/L. Similarly, Zhou et al. (2020) used evolutionary design and a biosensor to optimize (2S)-naringenin production in *E. coli*. These strategies resulted in high titers of (2S)-naringenin, reaching 588 mg/L.

In addition to *E. coli, S. cerevisiae* has also been engineered for naringenin production. Zhou et al. (2020) reported titers as high as 648.63 mg/L in *S. cerevisiae*. Several optimization strategies have been employed, including the use of evolutionary design algorithms for gene deletions and the development of biosensors to respond to environmental changes (Zhou et al., 2020; Fowler et al., 2009). *Corynebacterium glutamicum* expressing chalcone synthase and chalcone isomerase from *Petunia hybrida* demonstrated improved naringenin production, reaching 35 mg/L (Kallscheuer et al., 2016). *S. cerevisiae* strains overexpressing chalcone synthase and 4-coumarate CoA ligase from Alternaria sp. MG1 produced 18.5 mg/L of naringenin (Lu et al., 2021). Furthermore, Dunstan et al. (2020) successfully developed recombinant *E. coli* strains capable of producing 396–484 mg/L of naringenin.

### *1.3.6 Quercetin*

Flavonols, including quercetin, kaempferol, fisetin, and myricetin, belong to an important subfamily of flavonoids. Among them, quercetin is the most commonly found dietary flavonoid, and it can be obtained from sources such as onions, blueberries, cabbages, and peaches (Manach et al., 2004). Quercetin is renowned for its nutraceutical properties, particularly its anticancer effects. Studies have shown that quercetin modulates various pathways associated with inflammation and carcinogenesis in chemically induced colon cancer models in rodents (Russo et al., 2014).

Heterologous expression of quercetin has been achieved in *Streptomyces albus* and *Streptomyces coelicolor* (Marin et al., 2018). A combinatorial biosynthetic approach was employed to reconstruct the flavonoid pathway using a high-copy-number shuttle vector called pQR. This vector contained all seven genes necessary for quercetin biosynthesis, including tyrosine ammonia lyase from *Rhodobacter capsulatus*, 4-coumarate CoA ligase from *Streptomyces coelicolor*, chalcone synthase and chalcone isomerase from *Glycine max*, N3DOX from *Petroselinum crispum*, and flavonol synthase 1 and F3'H from *Arabidopsis thaliana*. The final vector was transformed and expressed in *S. albus*, resulting in low yields of 0.1 mg/L of quercetin. This study marked the first report of de novo biosynthesis of quercetin without the addition of precursor molecules to the culture medium. In *S. cerevisiae*, an enhanced titer of 0.38 mg/L of quercetin was achieved by overexpressing the flavanone 3-hydroxylase gene, which catalyzes the hydroxylation of kaempferol at the B3 position to yield quercetin. Naringenin was supplied as a precursor in this approach (Marin et al., 2018).

To enhance the production of flavonoids, one approach involves increasing the intracellular pool of malonyl CoA. This can be achieved by adding cerulenin, a fatty acid synthesis inhibitor, which elevates the levels of malonyl CoA (van Summeren-Wesenhagen & Marienhagen, 2015). Overexpression of acetyl CoA carboxylase is another strategy to enhance the pool of malonyl CoA. However, due to the metabolic burdens associated with gene overexpression, a more advanced approach utilizing CRISPR interference (CRISPRi) has been successfully employed in *E. coli* for improved flavonoid production. The CRISPRi system allows fine-tuning of genes involved in central metabolic pathways, enabling control over metabolic flux and identification of target genes that enhance malonyl CoA production (Gao et al., 2018).

Flavonols, which include quercetin, are derivatives of flavones. Flavonols are commonly found in tea, red wines, fruits, and vegetables (Panche et al., 2016). These compounds have diverse applications in the pharmaceutical industry, with quercetin exhibiting inhibitory effects on the metastasis of melanoma cells (Caltagirone et al., 2000). Biosynthesis of quercetin has been achieved by extending the kaempferol biosynthetic pathway through the overexpression of F3’H in *S. cerevisiae*. When naringenin was added to the cultures, the best engineered host strain produced 0.38 mg/L of quercetin during a 70-h cultivation period (Trantas et al., 2009). Recently, de novo biosynthesis of quercetin in actinomycetes was reported. In this study, an artificial pathway containing various enzymes from different sources, including TAL from *Rhodobacter capsulatus*, 4CL from *S. coelicolor*, CHS and CHI from *G. max*, naringenin 3-dioxygenase from *P. crispum*, and FLS and F3’H from *A. thaliana*, was constructed and introduced into *S. albus* and *S. coelicolor*. The recombinant *S. albus* strain achieved a higher titer of quercetin at 0.1 mg/L (Marín et al., 2018).

Fisetin, a compound structurally similar to quercetin, can be synthesized by hydroxylating fisetin at the C5 site to produce quercetin. A novel biosynthetic pathway for fisetin production was developed by utilizing liquiritigenin as an intermediate. De novo biosynthesis of kaempferol, quercetin, and fisetin from a simple carbon source in *S. cerevisiae* was achieved. The best engineered host strain produced 26.6 mg/L of kaempferol, 20.4 mg/L of quercetin, and 2.3 mg/L of fisetin directly from glucose (Marín et al., 2018). Strategies for enhancing flavonoid production involve increasing the pool of malonyl CoA, employing CRISPR interference for fine-tuning metabolic pathways, and constructing artificial biosynthetic pathways in microbial hosts. These approaches have been successful in improving the production of flavonoids such as quercetin and fisetin.

### 1.3.7 Carotenoids

Carotenoids are natural pigments that are lipid-soluble and highly valued for their colorant properties and antioxidant effects. They have nutraceutical benefits and can help prevent oxidative damage- and inflammation-related diseases such as cancer, cardiovascular diseases, atherosclerosis, neurodegenerative disorders, and diabetes (Ciriminna et al., 2016; Devraj et al., 2008; Rao & Rao, 2007). The global market for carotenoids is expected to reach $2.0 billion by 2022 (McWilliams, 2018).

Extensive research has focused on heterologous production of valuable carotenoids, such as β-carotene, lycopene, and astaxanthin, in noncarotenogenic microorganisms (Alper et al., 2005). Carotenoids are derived from two building blocks, isopentenyl pyrophosphate (IPP) and dimethylallyl pyrophosphate (DMAPP), which can be synthesized through either the mevalonic acid (MVA) pathway or the methylerythritol 4-phosphate (MEP) pathway. In *E. coli*, engineering efforts have led to the production of 3.2 g/L of carotenoids in a glycerol fed-batch experiment. This was achieved by improving the MEP pathway through the expression of isopentenyl pyrophosphate isomerase (FNI) from *B. subtilis* and geranyl diphosphate synthase (GPPS2) from *A. grandis*, while using glycerol as a carbon source to reduce acetic acid accumulation (Yang & Guo, 2014).

Another approach in *E. coli* optimized metabolic modules involved in β-carotene synthesis (CrtEXYIB from *P. agglomerans*), MEP pathway (overexpression of Dxs and Idi), as well as the TCA cycle and pentose phosphate pathway (PPP). This strategy enabled the production of 2.1 g/L of β-carotene in a fed-batch fermentation. The study also highlighted the importance of increasing NADPH supply for improving carotenoid production, which was found to be more crucial than ATP (Zhao et al., 2013). While bacteria have shown success in carotenoid production, recent studies have focused on the interaction between yeast and their ability to store molecules like carotenoids in lipid droplets. In *Y. lipolytica*, fine-tuning the expression of enzymes involved in carotenoid production, both native enzymes for increasing the supply of geranylgeranyl diphosphate (GGPP) precursor and exogenous enzymes carRP and carB from *Mucor circinelloides* for the conversion of GGPP into β-carotene, led to a strain capable of producing 4 g/L of β-carotene in a fed-batch fermentation process using a nitrogen-limited medium (Gao et al., 2017). Another engineered *Y. lipolytica* strain achieved even higher production, producing 6.5 g/L of β-carotene with a content of 89.6 mg/g dry cell weight in a fed-batch fermentation. This was achieved by rewiring the flux toward acetyl-CoA, along with overexpression of downstream geranylgeranyl diphosphate synthase (GGS1), *M. circinelloides* carRP, and carB under the control of a strong TEF promoter (Larroude et al., 2018).

Lycopene is another carotenoid of significant interest, and recent advancements have been made in the engineering of *S. cerevisiae* to enhance lycopene production (Ma et al., 2018). This engineering approach involved several metabolic modifications, including the overexpression of enzymes to redirect carbon flux from ethanol to acetyl-CoA and mevalonate precursors, increased generation of the cofactor

NADPH, and integration of the lycopene biosynthetic pathway (crtE from T. × media, crtB from *P. agglomerans*, and crtI from *B. trispora*). Optimization of triacylglycerol (TAG) metabolism was also achieved by overexpressing a fatty acid desaturase OLE1 to increase the supply of unsaturated fatty acids, and by deleting the Seipin complex gene FLD1 to regulate lipid-droplet size. This optimized strain produced 2.37 g/L of lycopene in a two-stage fed-batch fermentation process, with the first stage focused on biomass accumulation and the second stage for lycopene production from fed ethanol. Notably, this study achieved the highest lycopene content (73.3 mg/g dry cell weight) reported to date in oleaginous microbes or in *S. cerevisiae*, suggesting its potential as an industrial producer, potentially replacing the natural producer *B. trispora* (Ma et al., 2018).

In *E. coli*, lycopene production can be enhanced by overexpressing two major rate-limiting native enzymes, DXP synthase (Dxs) and IPP isomerase (Idi), to increase the supply of IPP and DMAPP. This leads to increased production of farnesyl diphosphate (FPP) by GPP/FPP synthase (IspA) (Alper et al., 2005). In contrast, in eukaryotic systems, FPP is synthesized through the native MVA pathway (Mata-Gómez et al., 2014). The FPP is then converted into lycopene by introducing a heterologous pathway containing geranylgeranyl pyrophosphate synthase (CrtE), phytoene synthase (CrtB), and phytoene desaturase (CrtI). An alternative strategy for carotenoid overproduction in *E. coli* involves the introduction of a heterologous MVA pathway. Through targeted engineering, a highly efficient lycopene-producing *E. coli* strain was constructed by introducing extra copies of the idi gene to leverage the MVA pathway and integrating the lycopene pathway. This strain achieved a lycopene titer of 1.44 g/L from glycerol in a fed-batch fermentation and was successfully scaled up to a 100 L fed-batch fermentation, reaching a titer of 1.23 g/L (Zhu et al., 2015).

### *1.3.8 β-Alanine*

β-Alanine, a naturally occurring β-type amino acid, has important physiological functions and is widely used as a precursor for various industrial chemicals in fields such as medicine, feed, food, and environmental applications. The biochemistry of β-alanine involves the action of l-aspartate-α-decarboxylase (PanD or ADC), which catalyzes the decarboxylation of l-aspartate to form β-alanine. Although enzymatic conversion of l-aspartate via ADC has been reported with high efficiency, it is an expensive method requiring large quantities of both the precursor l-aspartate and the enzyme (Shen et al., 2014). Therefore, there is great potential in de novo microbial production of β-alanine from inexpensive carbon sources.

Current efforts to increase β-alanine production primarily focus on bacterial hosts since fungal organisms like *S. cerevisiae* lack orthologs of panD and innate high fluxes toward the required precursors (Borodina et al., 2015). In *E. coli*, a novel metabolic pathway has been designed for β-alanine production from glucose, resulting in the highest reported titer to date of 32.3 g/L (Song et al., 2015). This strain overexpressed *C. glutamicum* PanD (which has higher specific activity than the *E. coli* counterpart), aspartase (AspA), and phosphoenolpyruvate carboxylase (Ppc) in a highly fumaric acid-producing *E. coli* background.

Another engineered *E. coli* strain was developed for β-alanine production through a series of genetic manipulations. The L-aspartate-α-decarboxylase gene, BtADC, from *Bacillus tequilensis* with higher catalytic activity was constitutively expressed in *E. coli*, leading to an increased production of β-alanine up to 2.76 g/L. Additionally, three native aspartate kinase genes (akI, akII, and akIII) were knocked out to prevent bypass loss of aspartate, resulting in a higher concentration of β-alanine (4.43 g/L). To further increase aspartate availability, the native AspC gene was replaced with PaeAspDH, an L-aspartate dehydrogenase gene from *Pseudomonas aeruginosa*, and the native AspA gene was overexpressed, leading to a further improvement in β-alanine production (9.27 g/L). To enhance biosynthesis of oxaloacetic acid (OAA), a combination of strategies including overexpression of native PPC, introduction of CgPC (a pyruvate decarboxylase from *Corynebacterium glutamicum*), and deletion of ldhA, pflB, pta, and adhE genes was employed, resulting in an engineered *E. coli* strain capable of producing 43.12 g/L β-alanine in fed-batch fermentation (Zou et al., 2020). These studies demonstrate the successful engineering of microbial strains, particularly *E. coli*, for the efficient production of β-alanine from inexpensive carbon sources, providing potential applications in various industrial sectors.

### 1.3.9 Polyunsaturated Fats (PUFAs)

Polyunsaturated fatty acids (PUFAs) are essential for the proper functioning of the body, but they are not synthesized internally and must be obtained from the diet. PUFAs can be categorized as omega-3 or omega-6 fatty acids, with important examples being α-linolenic acid (ALA), eicosapentaenoic acid (EPA), and docosahexaenoic acid (DHA). EPA and DHA are derived from ALA as a precursor molecule. Recent advancements in omics technology have opened new avenues for improving strains through metabolic engineering to produce nutraceuticals, including essential PUFAs, using engineered microorganisms in a cost-effective manner. Metabolic engineering allows the manipulation of biosynthetic pathways and optimization of microorganisms for substantial production of nutraceuticals, such as PUFAs, using simple carbon sources.

Certain strains of *Escherichia coli* and food-grade *Saccharomyces cerevisiae*, known as generally recognized as safe (GRAS) microorganisms, have been economically employed as “cell factories” for the production of desired nutraceuticals. In addition to these favored microorganisms, versatile traditional GRAS microorganisms like the oleaginous yeast *Yarrowia lipolytica* are also being utilized for nutraceutical biosynthesis (Darvishi et al., 2018). EPA and DHA are synthesized through the desaturase/elongase pathway, which is aerobic, although assembly occurs via the anaerobic polyketide synthase (PKS) pathway. Researchers from

DuPont introduced these biosynthetic genes into *Y. lipolytica*, resulting in a strain capable of synthesizing EPA at a rate of 56% of the total unsaturated fatty acids and approximately 15% of the dry cell weight (Xue et al., 2013). Another engineered industrial strain, *Y. lipolytica* Z5567, was developed by the same group to direct carbon flux specifically towards EPA biosynthesis by eliminating β-oxidation, resulting in EPA production at 25% of the dry cell weight (Xie et al., 2015). Zhu and Jackson developed a two-stage fed-batch bioprocess system that utilized nitrogen-rich media for growth and nitrogen-limiting conditions for oil biosynthesis, resulting in EPA production at 25% of the dry cell weight (Zhu & Jackson, 2015).

Using this technology for PUFA production, two commercial products, New Harvest™ EPA oil and Verlasso® sustainably farmed salmon, have been developed (Zhu and Jackson, 2015). Additionally, there are various alternative hosts and technologies for DHA production, including several marine microalga strains that naturally have high DHA content (30–40% of total fatty acids) (Adarme-Vega et al., 2012). However, traditional hosts like bacteria and yeast often have limitations in DHA production (less than 6% of total fatty acids) (Damude et al., 2016; Peng et al., 2016), so their production in these hosts is not extensively discussed. Another omega-3 fatty acid, α-linolenic acid (ALA), has been explored in *Y. lipolytica* as well. ALA biosynthesis requires a Δ15-desaturase to convert native unsaturated fatty acids like oleic acid (C18:1n-9) and linoleic acid (C18:2n-6) into ALA (Cui et al., 2016). By using a previously engineered strain of *Y. lipolytica* that produces nearly 80% of lipids as unsaturated C18 fatty acids (Liu et al., 2015), a platform for ALA biosynthesis was created. Heterologous expression of a codon-optimized, bifunctional Δ12/Δ15-desaturase from *Rhodosporidium kratochvilovae*, coupled with low-temperature fermentation at 20 °C, significantly increased ALA content. The resulting strain was capable of producing ALA up to 30% of total fatty acids and achieving titers of 1.4 g/L ALA in fed-batch fermentation, the highest reported titer in a yeast host (Cordova & Alper, 2018). These findings demonstrate the utility of microorganisms, especially oleaginous yeasts, in the production of nutritional fatty acids.

### *1.3.10 Prebiotics*

A prebiotic is defined by the Food and Agriculture Organization of the United Nations (FAO) as "a nonviable food component that confers a health benefit on the host associated with modulation of the microbiota" (Pineiro et al., 2008). Prebiotics are non-digestible saccharide polymers composed of 3–10 monomeric sugar units. The beneficial effects of prebiotics on humans stem from their ability to modulate the diversity and metabolic activity of the gut microbiota, which in turn can have a significant impact on the host immune system. Consequently, prebiotics have the potential to be used in the treatment of inflammatory diseases by promoting the growth of probiotic bacteria such as Bifidobacteria or Lactobacillus species within the gut microbiota (Lin et al., 2014).

Representative prebiotics include inulin, fructo-oligosaccharides (FOS), and galacto-oligosaccharides (GOS). In *Lactococcus lactis*, the expression of codon-optimized β-galactosidase from the hyperthermophile *Sulfolobus solfataricus* efficiently converts lactose to GOS Yu & O' Sulllivan, (2014). Another prebiotic, 2′-fucosyllactose (2'-FL), is one of the most abundant oligosaccharides in human milk and has been approved as a nutritional additive in infant and toddler formulas. By increasing the availability of GDP-L-fucose and overexpressing fucosyltransferase, 2'-FL can be produced from lactose and glycerol in *Escherichia coli* (Lee et al., 2012; Baumgartner et al., 2013). Inulin and inulin-type fructans, which are soluble dietary fibers, can be produced by probiotic *Lactobacillus gasseri* strains (Anwar et al., 2010). GOS, lactose-derived galactose polymers with shorter polymer chains, can be produced by *Kluyveromyces lactis* with a high titer from lactose (Rodriguez-Colinas et al., 2011).

Researchers have also explored the production of prebiotics using enzymatic reactions. For example, the production of inulooligosaccharides was achieved by heteroexpression of an endoinulinase-encoding gene from *Aspergillus ficuum* in *Escherichia coli* BL21. Optimization of reaction conditions resulted in a high yield of inulooligosaccharides (Wang et al., 2016). Additionally, the combined use of levansucrase from *Bacillus amyloliquefaciens* and endo-inulinase from *Aspergillus niger* in a one-step reaction using sucrose as the substrate has been investigated for the production of fructooligosaccharides (FOS) and oligolevans (Tian et al., 2014). Furthermore, solid-state fermentation (SSF) systems have been employed for prebiotic production. An augmented simplex lattice design was used to optimize a three-component mixture for enzyme production and FOS production, demonstrating the potential for obtaining large amounts of enzymes and FOS using low-cost substrates (Sathish & Prakasham et al., 2013). Fed-batch synthesis of galactooligosaccharides from lactose with β-galactosidase from *Aspergillus oryzae* was experimentally evaluated, and optimal control techniques were employed to maximize the reaction yield (Vera et al., 2014). Several microorganisms, including molds, yeasts, and bacteria, have been assessed for their ability to produce fructosyltransferase and generate FOS from sucrose, with strains such as *A. flavus* (NFCCI 2364), *A. niger* (SI 19), and *Penicillium islandicum* (MTCC 4926) identified as promising candidates for industrial-scale production (Ganaie et al., 2013).

### *1.3.11 Polysaccharides*

Polysaccharides, which are composed of monosaccharide building blocks, are the primary constituents of carbohydrates in nature. They can be produced by various bacteria, yeast, fungi, or extracted from plant and animal tissues. Microbial polysaccharides have gained recognition as a source of nutraceuticals due to their diverse properties, including antitumor, antimicrobial, immune-regulating, and hypoglycemic effects. Natural polysaccharides have been found to enhance the immune system and have potential therapeutic applications in various diseases. Additionally, polysaccharides find extensive use in the food and dairy industries (Mohammed et al., 2021; Shakya, 2020).

Bacterial polysaccharides such as xanthan, gellan, dextrans, and alginate can be commercially produced through microbial production and purification. The production of polysaccharides involves the biosynthesis of energy-rich monosaccharide precursors, primarily nucleoside diphosphate sugars (NDP-sugars) derived from phosphorylated sugars (Mandakovic et al., 2020). Exopolysaccharides (EPS) are primarily produced through metabolic engineering of Lactococcus and Streptococcus for their use in dairy products. Fungal polysaccharides with bioactive properties have shown extensive benefits such as immunostimulation, antitumor, antimicrobial, antioxidant, hypocholesterolemic, and hypoglycemic effects, making them promising for both nutraceutical and pharmaceutical applications (Giavasis, 2014). EPS secretion in bacteria involves complex mechanisms to transport high molecular weight hydrophilic polymers across the cell membrane without compromising the membrane's integrity (Vaishnav et al., 2021).

In gram-negative bacteria, which exhibit a wide range of EPS molecular structures, two identified mechanisms are involved in biosynthesis and release (Mohd Nadzir et al., 2021). These mechanisms include the Wzx flippase, Wzy polymerase, and Wzz chain-length regulator proteins located in the integral inner membrane. However, detailed structural and functional characterization of these components remains a challenge (Wiseman et al., 2021). The polymer repeat complex is formed and polymerized in the periplasm, mainly on the inner surface of the cytoplasmic membrane. The transporter-dependent phase of ABC occurs on the cytoplasmic surface of the inner membrane, facilitating polymerization (Soumya & Nampoothiri, 2021). Rather than extracting polysaccharides from animal tissues, certain animal polysaccharides such as hyaluronic acid (HA), chondroitin, and heparosan can be produced using microbial hosts. Microbial production of hyaluronic acid has been achieved in various hosts, including *E. coli*, *Lactococcus lactis*, and *Streptomyces albulus* (Yoshimura et al., 2015; Yu & Stephanopoulos, 2008). Engineered *E. coli* has been used to produce medicinally important polysaccharides such as heparosan and chondroitin, reaching relatively high titers of 1.88 g/L and 2.4 g/L, respectively (He et al., 2015). The extracellular glucan scleroglucan, produced by the mycelia of *Sclerotium rolfsii*, is a potential antitumor and antiviral polysaccharide whose yield can be enhanced by the addition of L-lysine and uridine mono-phosphate (UMP) (Giavasis, 2014).

### *1.3.12 Poly Amino Acids*

Polyamino acid biosynthesis is an intriguing example of biopolymer synthesis that occurs through the fermentation process. Polyamino acids have a wide range of applications, including food additives, biomedical agents, and as biodegradable and renewable resources. These materials are environmentally friendly, biodegradable, and not reliant on oil-based resources. In the literature, three commonly studied natural polyamino acids are poly-ε-lysine, poly-γ-glutamic acid, and cyanophycin.

Polyamino acids are synthesized in microorganisms through ribosome-independent enzymatic processes, which distinguish them from protein synthesis. The three types of polyamino acids found in nature are poly-γ-glutamic acid (γ-PGA), poly-ε-L-lysine (ε-PL), and multi-L-arginyl-poly(L-aspartic acid) (cyanophycin). γ-PGA is a water-soluble and biodegradable polymer that can be used as a degradable biomaterial, such as drug carriers or hydrogels. Bacillus species can produce large quantities (20–50 g/L) of γ-PGA when supplied with L-glutamic acid. Co-cultivation of *Corynebacterium glutamicum* and *Bacillus subtilis*, optimizing glucose as the carbon source, can achieve a maximum γ-PGA production of 32.8 g/L (Shi et al., 2007). Strain M-Z18 reached 35.14 g/L of γ-PGA when using glucose and glycerol as carbon sources (Chen et al., 2012). A novel strain with high γ-PGA productivity, named *Bacillus subtilis* ZJU-7, was isolated from fermented bean curd, a traditional Chinese food. The most suitable carbon and nitrogen sources for this strain were sucrose and tryptone, respectively. The addition of exogenous L-glutamic acid enhanced γ-PGA production, and after 24 h of cultivation at 37 °C, the yield of γ-PGA reached 54.4 g/L, the highest reported productivity (Shi et al., 2007).

Cyanophycin is produced by cyanobacteria and certain chemotrophic bacteria, such as *Acinetobacter calcoaceticus*, from equimolar amounts of arginine and aspartic acid. It can be used as a precursor for dipeptides in nutritional and therapeutic applications. Overexpression of cyanophycin synthetase cphA from Synechocystis sp. PCC 6803 in *E. coli* achieved cyanophycin production with a productivity of 120 mg/g CDW (Tseng et al., 2012a, b). ε-PL is a microbial-derived homo-poly amino acid synthesized by polymerization of lysine via ε-PL synthetase (PLS). It possesses antibacterial and anticancer activities and has been approved as a food preservative or dietary agent in Japan and the United States. Streptomyces species, such as *Streptomyces albulus*, are known producers of ε-PL. In Japan, ε-PL has been commercialized using a mutant strain derived from *Streptomyces albulus* 346 (Shima & Sakai, 1977). Initially, wild-type *Streptomyces albulus* 346 achieved a maximum ε-PL productivity of only about 0.5 g/L under optimized culture conditions. However, under acidic pH conditions, ε-PL accumulation significantly improved, reaching a maximum productivity of 4–5 g/L. *S. albulus* 410 (S410), a high ε-PL producer, was used to evaluate the effect of pH value on ε-PL productivity, with pH control fermentation enhancing ε-PL productivity from 5.7 to 48.3 g/L in a batch culture (Kahar et al., 2001). Glucose concentration in the broth was controlled at 10 g/L to prevent pH increase, which could lead to the degradation of produced ε-PL.

## 1.4 Challenges and Future Prospects of Commercial Production of Nutraceuticals

Microbial sources are particularly advantageous for the production of nutraceuticals due to their scalability and ease of manipulation. Advancements in strain development, systems biology, metabolic engineering, and protein engineering have significantly improved the quality and quantity of these compounds. Efficient fermentations offer predictable yields and are not affected by external factors such as climate or environment. However, further research is needed to optimize bioactive compound characteristics, including composition and yield, by identifying the optimal growth parameters, utilizing genetically modified organisms to enhance production, and employing various elicitors for nutraceutical production (Das et al., 2007).

Metabolic manipulation technology has played a significant role in enhancing nutraceutical production in microbes. The introduction of novel enzymes, screening of gene sources, protein manipulation, and regulator engineering have been successfully implemented in various metabolic components to enhance nutraceutical synthesis. Although there are genetically and metabolically engineered microbial production strains available and utilized in the industry, more efforts are required to achieve microbial overproduction of desired compounds. Metabolic engineering enables improvements in product yields, transfer of pathways from slow-growing organisms to faster-growing ones, and directed biosynthesis of pigment analogs to modify color or other properties. The utilization of CRISPR-Cas9 and heterologous expression of biosynthetic pathways from known or novel producers can further facilitate the creation of cell factories (Sankari et al., 2018; Nielsen & Nielsen, 2017).

While metabolic engineering is valuable, it also presents regulatory challenges. Regulatory frameworks and safety considerations surrounding the use of genetically modified organisms (GMOs) for nutraceutical production can pose challenges. Compliance with regulatory guidelines and addressing public concerns related to GMOs is necessary for the commercialization of microbial-based nutraceutical production. Scaling up microbial production from laboratory to industrial scale poses challenges related to fermentation processes. Factors such as oxygen transfer, nutrient supply, pH control, and removal of inhibitory by-products need to be carefully managed to ensure optimal growth and productivity of microbial cultures. Further, extracting, purifying, and concentrating the nutraceutical compounds from the microbial cultures can be challenging and costly. The genetic stability of bacterial or yeast strains limits their industrial application. The cost of microbial production processes needs to be competitive with other production methods to make microbial-based nutraceuticals commercially viable. Optimizing production strategies, reducing raw material costs, and improving overall process efficiency are ongoing challenges in achieving cost competitiveness.

Currently, only a few microbial processes have been developed for chromosomal genes and their integration regulation, which is unstable and relies on antibiotics, resulting in a high metabolic load (Furubayashi et al., 2015). To address these challenges, various technologies have been employed, such as marker-free genetic integration techniques based on CRISPR-Cas, highly stable heterologous gene expression technologies based on incoherent feedforward loops, and antibiotics-free plasmids based on stable and tunable gene expression strategies (Klompe et al., 2019; Meng et al., 2020; Segall-Shapiro et al., 2018; Pieters et al., 2021; Kang et al., 2018; Egger et al., 2020). Additionally, transcription factor-based or riboswitch-based biosensors have been developed to detect specific intracellular or extracellular metabolites for active metabolic network regulation. However, there is a need for an optimized experimental model to effectively apply these elements in the construction of engineered microbes. Ongoing research focuses on the design, construction, and validation of programmed gene circuits and elements to further enhance microbial engineering (Bashor et al., 2019). Advancements in analytical technologies have also contributed to strain construction and product discovery. For example, optically guided matrix-assisted laser desorption/ionization MS/MS has been used for high-throughput analysis of microbial colony profiling. Continuous developments in system and synthetic biology have improved the engineering tools and engineered microbes to meet the requirements of current and future nutraceutical compounds (Das et al., 2007).

## 1.5 Conclusion

Microbial metabolic modification is a promising approach for producing nutraceutical compounds, offering an alternative to chemical synthesis and extraction methods. This method allows for the synthesis of clean compounds without the need for expensive equipment. Over the past few decades, significant progress has been made in the field of nutraceutical production through metabolic engineering of microbial platforms. This advancement addresses the growing demand for value-added nutraceuticals to enhance human health and combat diseases. By bypassing the challenges associated with isolating compounds directly from plants and animals, microbial metabolic engineering provides a sustainable and environmentally friendly technology for commercial synthesis. As the demand for these compounds continues to increase with population growth, the application of metabolic engineering becomes increasingly important. Microbial metabolic engineering enables not only laboratory-scale production but also industrial-scale manufacturing of complex natural products, including carotenoids, using simple carbon sources. With the elucidation of biosynthetic pathways and the development of synthetic biology strategies, it becomes possible to produce novel or more complex value-added compounds in microbes using naturally occurring processes. Advances in omics technologies, synthetic biology, and genome engineering have expanded the range of heterologous and homologous microbial hosts available, ensuring the ability to meet the requirements for current and future nutraceutical compounds.

## References

Adarme-Vega, T. C., Lim, D. K. Y., Timmins, M., Vernen, F., Li, Y., & Schenk, P. M. (2012). Microalgal biofactories: A promising approach towards sustainable omega-3 fatty acid production. *Microbial Cell Factories, 11*, 96. https://doi.org/10.1186/1475-2859-11-96

Afonso, M. S., Ferreira, S., Domingues, F. C., & Silva, F. (2014). Resveratrol production in bioreactor: Assessment of cell physiological states and plasmid segregational stability. *Biotechnology Reports, 5*, 7–13.

Alper, H., Miyaoku, K., & Stephanopoulos, G. (2005). Construction of lycopene-overproducing *E. coli* strains by combining systematic and combinatorial gene knockout targets. *Nature Biotechnology, 23*(5), 612. https://doi.org/10.1038/nbt1083

Amiri-Jami, M., & Griffiths, M. W. (2010). Recombinant production of omega-3 fatty acids in Escherichia coli using a gene cluster isolated from Shewanella baltica MAC1. *Journal of Applied Microbiology, 109*, 1897–1905.

Anwar, M. A., Kralj, S., Piqué, A. V., Leemhuis, H., van der Maarel, M. J., & Dijkhuizen, L. (2010). Inulin and Levan synthesis by probiotic lactobacillus gasseri strains: Characterization of three novel fructansucrase enzymes and their fructan products. *Microbiology, 156*, 1264–1274. https://doi.org/10.1099/mic.0.036558-0

Bashor, C. J., Horwitz, A. A., Peisajovich, S. G., Lim, W. A., & Voigt, C. A. (2019). Rewiring cells: Synthetic biology as a tool to interrogate the organizational principles of living systems. *Annual Review of Biophysics, 48*, 21–39.

Baumgärtner, F., Seitz, L., Sprenger, G. A., & Albermann, C. (2013). Construction of Escherichia coli strains with chromosomally integrated expression cassettes for the synthesis of 2′-fucosyllactose. *Microbial Cell Factories, 12*, 40. https://doi.org/10.1186/1475-2859-12-40

Beekwilder, J., Wolswinkel, R., Jonker, H., Hall, R., de Vos, C. R., & Bovy, A. (2006). Production of resveratrol in recombinant microorganisms. *Applied and Environmental Microbiology, 72*, 5670–5672.

Bhan, N., Li, L., Cai, C., Xu, P., Linhardt, R. J., & Koffas, M. A. (2015). Enzymatic formation of a resorcylic acid by creating a structure-guided single-point mutation in stilbene synthase. *Protein Science, 24*, 167–173.

Borodina, I., Kildegaard, K. R., Jensen, N. B., Blicher, T. H., Maury, J., Sherstyk, S. & Nielsen, J. (2015). Establishing a synthetic pathway for high-level production of 3-hydroxypropionic acid in Saccharomyces cerevisiae via β-alanine. *Metabolic engineering, 27*, 57–64.

Braga, A., Oliveira, J., Silva, R., Ferreira, P., Rocha, I., Kallscheuer, N., Marienhagen, J., & Faria, N. (2018). Impact of the cultivation strategy on resveratrol production from glucose in engineered Corynebacterium glutamicum. *Journal of Biotechnology, 265*, 70–75.

Brown, S., Clastre, M., Courdavault, V., & O'Connor, S. E. (2015). De novo production of the plant-derived alkaloid strictosidine in yeast. *Proceedings of the National Academy of Sciences, 112*(10), 3205–3210.

Caltagirone, S., Rossi, C., Poggi, A., Ranelletti, F. O., Natali, P. G., Brunetti, M., et al. (2000). Flavonoids apigenin and quercetin inhibit melanoma growth and metastatic potential. *International Journal of Cancer, 87*(4), 595–600.

Camacho-Zaragoza, J. M., Hernández-Chávez, G., Moreno-Avitia, F., Ramírez- Iñiguez, R., Martínez, A., Bolívar, F., & Gosset, G. (2016). Engineering of a microbial coculture of Escherichia coli strains for the biosynthesis of resveratrol. *Microbial Cell Factories, 15*, 163.

Chen, X.-S., Ren, X.-D., Dong, N., Li, S., Li, F., Zhao, F.-L., Tang, L., Zhang, J.-H., & Mao, Z.-G. (2012). Culture medium containing glucose and glycerol as a mixed carbon source improves ε-poly-L-lysine production by Streptomyces sp. M-Z18. *Bioprocess and Biosystems Engineering, 35*, 469–475.

Chen, Y., Chen, Y., Shi, Y., & Huang, L. (2012b). Biosynthesis of poly(γ-glutamic acid) from glucose by recombinant Bacillus subtilis. *Biotechnology Letters, 34*(8), 1453–1458.

Choi, J. W., Yim, S. S., Lee, S. H., Kang, T. J., Park, S. J., & Jeong, K. J. (2015). Enhanced production of gamma-aminobutyrate (GABA) in recombinant Corynebacterium glutamicum by expressing glutamate decarboxylase active in expanded pH range. *Microbial Cell Factories, 14*, 21.

Ciriminna, R., Fidalgo, A., Meneguzzo, F., Ilharco, L. M., & Pagliaro, M. (2016). Lycopene: Emerging production methods and applications of a valued carotenoid. *ACS Sustainable Chemistry & Engineering, 4*(2), 643–650. https://doi.org/10.1021/acssuschemeng.5b01233

Cordova, L. T., & Alper, H. S. (2018). Production of α-linolenic acid in Yarrowia lipolytica using low-temperature fermentation. *Applied Microbiology and Biotechnology, 102*, 8809–8816. https://doi.org/10.1007/s00253-018-9349-y

Cui, J., He, S., Ji, X., Lin, L., Wei, Y., & Zhang, Q. (2016). Identification and characterization of a novel bifunctional Δ12/Δ15-fatty acid desaturase gene from Rhodosporidium kratochvilovae. *Biotechnology Letters, 38*, 1155–1164. https://doi.org/10.1007/s10529-016-2079-1

Da Silva, S. B., Cantarelli, V. V., & Ayub, M. A. Z. (2014). Production and optimization of poly-γ-glutamic acid by Bacillus subtilis BL53 isolated from the Amazonian environment. *Bioprocess and Biosystems Engineering, 37*, 469–479.

Damude, H. G., Macool, D. J., Picataggio, S. K., Ragghianti, J. J., Seip, J. E., Xue, Z., Yadav, N. S., Zhang, H., & Zhu, Q. (2016). Docosahexaenoic acid producing strains of Yarrowia lipolytica. US Patent 2009, US7,550,286.

Darvishi, F., Ariana, M., Marella, E. R., & Borodina, I. (2018). Advances in synthetic biology of oleaginous yeast Yarrowia lipolytica for producing non-native chemicals. *Applied Microbiology and Biotechnology, 102*, 5925–5938. https://doi.org/10.1007/s00253-018-9099-x

Das, A., Yoon, S. H., Lee, S. H., Kim, J. Y., Oh, D. K., & Kim, S. W. (2007). An update on microbial carotenoid production: Application of recent metabolic engineering tools. *Applied Microbiology and Biotechnology, 77*, 505–512. https://doi.org/10.1007/s00253-007-1206-3

De Luca, V., Salim, V., Atsumi, S. M., & Yu, F. (2012). Mining the biodiversity of plants: A revolution in the making. *Science, 336*, 1658–1661.

Devaraj, S., Mathur, S., Basu, A., Aung, H. H., Vasu, V. T., Meyers, S., et al. (2008). A dose-response study on the effects of purified lycopene supplementation on biomarkers of oxidative stress. *Journal of the American College of Nutrition, 27*(2), 267–273. https://doi.org/10.1080/07315724.2008.10719757

Dunstan, M. S., Robinson, C. J., Jervis, A. J., et al. (2020). Engineering Escherichia coli towards de novo production of gatekeeper (2S)-flavanones: Naringenin, pinocembrin, eriodictyol and homoeriodictyol. *Synthetic Biology*. https://doi.org/10.1093/synbio/ysaa012

Egger, S., Cha, H. J., & Park, H. (2020). Tuning gene expression in Escherichia coli for membrane protein overexpression. *FEMS Microbiology Letters, 367*(14), fnaa106.

Fowler, Z. L., Gikandi, W. W., & Koffas, M. A. G. (2009). Increased Malonyl coenzyme a biosynthesis by tuning the Escherichia coli metabolic network and its application to flavanone production. *Applied and Environmental Microbiology, 75*(18), 5831–5839.

Furubayashi, M., Ikeuchi, A., Kondo, A., & Soma, Y. (2015). Current status and potential of microbial cell factories for producing nutraceuticals. *Applied Microbiology and Biotechnology, 99*(4), 1577–1588.

Ganaie, M. A., Gupta, U. S., & Kango, N. (2013). Screening of biocatalysts for the transformation of sucrose to fructooligosaccharides. *Journal of Molecular Catalysis B: Enzymatic, 97*, 12–17. https://doi.org/10.1016/j.molcatb.2013.08.008

Ganesan, V., Li, Z., Wang, X., & Zhang, H. (2017). Heterologous biosynthesis of natural product naringenin by co-culture engineering. *Synthetic and Systems Biotechnology, 2*, 236–242. https://doi.org/10.1016/j.synbio.2017.08.003

Gao, S., Tong, Y., Zhu, L., Ge, M., Zhang, Y., Chen, D., Jiang, Y., & Yang, S. (2017). Iterative integration of multiple-copy pathway genes in Yarrowia lipolytica for heterologous beta-carotene production. *Metabolic Engineering, 41*, 192–201. https://doi.org/10.1016/j.ymben.2017.04.009

Gao, C., Wang, S., Hu, G., Guo, L., Chen, X., Xu, P., & Liu, L. (2018). Engineering Escherichia coli for malate production by integrating modular pathway characterization with CRISPRi-guided multiplexed metabolic tuning. *Biotechnology and Bioengineering, 115*(3), 661–672.

Giavasis, I. (2014). Bioactive fungal polysaccharides as potential functional ingredients in food and nutraceuticals. *Current Opinion in Biotechnology, 26*, 162–173. https://doi.org/10.1016/j.copbio.2013.12.004

Hawkins, K. M., & Smolke, C. D. (2008). Production of benzylisoquinoline alkaloids in Saccharomyces cerevisiae. *Nature Chemical Biology, 4*, 564–573.

He, W., Fu, L., Li, G., Jones, J. A., Linhardt, R. J., & Koffas, M. (2015). Production of chondroitin in metabolically engineered E. coli. *Metabolic Engineering, 27*, 92–100. https://doi.org/10.1016/j.ymben.2014.10.004

He, X.-Z., Li, W.-S., Blount, J. W., & Dixon, R. A. (2008). Regioselective synthesis of plant (iso) flavone glycosides in Escherichia coli. *Applied Microbiology and Biotechnology, 80*, 253–260.

Hwang, E. I., Kaneko, M., Ohnishi, Y., & Horinouchi, S. (2003). Production of plant-specific flavanones by Escherichia coli containing an artificial gene cluster. *Applied and Environmental Microbiology, 69*, 2699–2706.

Ibrahim, G. G., & Yan, Y. (2022). Microbial production of resveratrol. In S. M. Jafari & F. D. Harzevili (Eds.), *Microbial production of food bioactive compounds* (pp. 1–15). Springer.

Jain, N., & Ramawat, K. G. (2013). Nutraceuticals and antioxidants in prevention of diseases. In *Natural products* (pp. 2559–2580). Springer.

Jin, J., Wang, Y., Yao, M., Gu, X., Li, B., Liu, H., Ding, M., Xiao, W., & Yuan, Y. (2018). Astaxanthin overproduction in yeast by strain engineering and new gene target uncovering. *Biotechnology for Biofuels, 11*, 230. https://doi.org/10.1186/s13068-018-1226-5

Jones, J. A., Vernacchio, V. R., Sinkoe, A. L., Collins, S. M., Ibrahim, M. H. A., Lachance, D. M., Hahn, J., & Koffas, M. A. G. (2016). Experimental and computational optimization of an Escherichia coli co-culture for the efficient production of flavonoids. *Metabolic Engineering, 35*, 55–63.

Kahar, P., Iwata, T., Hiraki, J., Park, Y. E., & Okabe, M. (2001). Enhancement of ε-polylysine production by Streptomyces albulus strain 410 using pH control. *Journal of Bioscience and Bioengineering, 91*(2), 190–194. https://doi.org/10.1016/S1389-1723(01)80208-6

Kallscheuer, N., Vogt, M., Stenzel, A., Gätgens, J., Bott, M., & Marienhagen, J. (2016). Construction of a Corynebacterium glutamicum platform strain for the production of stilbenes and (2S)-flavanones. *Metabolic Engineering, 38*, 47–55.

Kang, S. G., Kim, S. K., & Park, S. J. (2018). Coordinated feedback regulation of recombinant protein expression by combined auto-induction and temperature-sensitive cell growth. *Applied Microbiology and Biotechnology, 102*(3), 1409–1417.

Katsuyama, Y., Miyahisa, I., Funa, N., & Horinouchi, S. (2007). One-pot synthesis of genistein from tyrosine by coincubation of genetically engineered Escherichia coli and Saccharomyces cerevisiae cells. *Applied Microbiology and Biotechnology, 73*, 1143–1149. https://doi.org/10.1007/s00253-006-0568-2

Katz, M., Smits, H. P., Förster, J., Nielsen, J. B., & Evola, S. A. (2015). Metabolically engineered cells for the production of resveratrol or an oligomeric or glycosidically-bound derivative thereof. (U.S. Patent No. 9,404,129).

Kildegaard, K. R., Adiego-Pérez, B., Doménech Belda, D., Khangura, J. K., Holkenbrink, C., & Borodina, I. (2017). Engineering of Yarrowia lipolytica for production of astaxanthin. *Synthetic and Systems Biotechnology, 2*, 287–294.

Kim, B.-G., Kim, H. J., & Ahn, J.-H. (2012). Production of bioactive flavonol rhamnosides by expression of plant genes in Escherichia coli. *Journal of Agricultural and Food Chemistry, 60*, 11143–11148.

Kim, E. M., Woo, H. M., Tian, T., Yilmaz, S., Javidpour, P., Keasling, J. D., & Kim, H. U. (2017). Autonomous control of metabolic state by a quorum sensing (QS)-mediated regulator for bisabolene production in engineered E. coli. *Metabolic Engineering, 44*, 325–336.

Kim, M.-J., Kim, B.-G., & Ahn, J.-H. (2013). Biosynthesis of bioactive O-methylated flavonoids in Escherichia coli. *Applied Microbiology and Biotechnology, 97*, 7195–7204.

Kim, S. Y., Lee, H. R., Park, K.-S., Kim, B.-G., & Ahn, J.-H. (2015). Metabolic engineering of Escherichia coli for the biosynthesis of flavonoid-O-glucuronides and flavonoid-O-galactoside. *Applied Microbiology and Biotechnology, 99*, 2233–2242.

Klompe, S. E., Vo, P. L. H., Halpin-Healy, T. S., Sternberg, S. H., & Beaumont, H. J. (2019). DNA modification-dependent activity and cell-to-cell variation of CRISPR-associated gene editors. *Nature Communications, 10*(1), 1–9.

Koksal, M., Chou, W. K., Cane, D. E., & Christianson, D. W. (2012). Structure of geranyl diphosphate C-methyltransferase from Streptomyces coelicolor and implications for the mechanism of isoprenoid modification. *Biochemistry, 51*(14), 3003–3010.

Koopman, F., Beekwilder, J., Crimi, B., van Houwelingen, A., Hall, R. D., Bosch, D., van Maris, A. J., Pronk, J. T., & Daran, J.-M. (2012). De novo production of the flavonoid naringenin in engineered Saccharomyces cerevisiae. *Microbial Cell Factories, 11*, 155.

Larroude, M., Celinska, E., Back, A., Thomas, S., Nicaud, J.-M., & Ledesma-Amaro, R. (2018). A synthetic biology approach to transform Yarrowia lipolytica into a competitive biotechnological producer of beta-carotene. *Biotechnology and Bioengineering, 115*, 464–472.

Lee, W.-H., Pathanibul, P., Quarterman, J., Jo, J.-H., Han, N. S., Miller, M. J., Jin, Y.-S., & Seo, J.-H. (2012). Whole cell biosynthesis of a functional oligosaccharide, 2′-fucosyllactose, using engineered Escherichia coli. *Microbial Cell Factories, 11*, 48. https://doi.org/10.1186/1475-2859-11-48

Lemuth, K., Steuer, K., & Albermann, C. (2011). Engineering of a plasmid-free Escherichia coli strain for improved in vivo biosynthesis of astaxanthin. *Microbial Cell Factories, 10*, 29.

Leonard, E., & Koffas, M. A. (2007). Engineering of artificial plant cytochrome P450 enzymes for synthesis of isoflavones by Escherichia coli. *Applied and Environmental Microbiology, 73*, 7246–7251.

Li, M., Kildegaard, K. R., Chen, Y., Rodriguez, A., Borodina, I., & Nielsen, J. (2015). De novo production of resveratrol from glucose or ethanol by engineered Saccharomyces cerevisiae. *Metabolic Engineering, 32*, 1–11.

Li, X.-R., Tian, G.-Q., Shen, H.-J., & Liu, J.-Z. (2015). Metabolic engineering of Escherichia coli to produce zeaxanthin. *Journal of Industrial Microbiology & Biotechnology, 42*, 627–636.

Li, M., Schneider, K., Kristensen, M., Borodina, I., & Nielsen, J. (2016). Engineering yeast for high-level production of stilbenoid antioxidants. *Scientific Reports, 6*, 36827.

Lin, C.-S., Chang, C.-J., Lu, C.-C., Martel, J., Ojcius, D. M., Ko, Y.-F., Young, J. D., & Lai, H.-C. (2014). Impact of the gut microbiota,.prebiotics, and probiotics on human health and disease. *Biomedical Journal, 37*, 259–268. https://doi.org/10.4103/2319-4170.138314

Liu, L., Markham, K., Blazeck, J., Zhou, N., Leon, D., Otoupal, P., & Alper, H. S. (2015). Surveying the lipogenesis landscape in Yarrowia lipolytica through understanding the function of a Mga2p regulatory protein mutant. *Metabolic Engineering, 31*, 102–111. https://doi.org/10.1016/j.ymben.2015.08.005

Lu, Y., Song, Y., Zhu, J., et al. (2021). Potential application of CHS and 4CL genes from grape endophytic fungus in production of naringenin and resveratrol and the improvement of polyphenol profiles and flavour of wine. *Food Chemistry, 347*, 128972. https://doi.org/10.1016/j.foodchem.2020.128972

Luo, Y., Li, B.-Z., Liu, D., Zhang, L., Chen, Y., Jia, B., & Zhang, Y. (2015). Engineered biosynthesis of natural products in heterologous hosts. *Chemical Society Reviews, 44*(15), 5265–5290.

Ma, T., Shi, B., Ye, Z., Li, X., Liu, M., Chen, Y., Xia, J., Nielsen, J., Deng, Z., & Liu, T. (2018). Lipid engineering combined with systematic metabolic engineering of Saccharomyces cerevisiae for high-yield production of lycopene. *Metabolic Engineering, 52*, 134–142. https://doi.org/10.1016/j.ymben.2018.10.001

Madhavan, A., Arun, K. B., Sindhu, R., Pandey, A., & Binod, P. (2019). Tailoring of microbes for the production of high-value plant-derived compounds: From pathway engineering to fermentative production. *Biochimica et Biophysica Acta (BBA): Proteins and Proteomics, 1867*(1), 140262.

Manach, C., Scalbert, A., Morand, C., et al. (2004). Polyphenols: Food sources and bioavailability. *The American Journal of Clinical Nutrition, 79*(5), 727–747. https://doi.org/10.1093/ajcn/79.5.727

Mandakovic, D., Cintolesi, Á., Maldonado, J., Mendoza, S. N., Aïte, M., Gaete, A., Saitua, F., Allende, M., Cambiazo, V., & Siegel, A. (2020). Genome-scale metabolic models of microbacterium species isolated from a high altitude desert environment. *Scientific Reports, 10*(1), 5560. https://doi.org/10.1038/s41598-020-62130-8

Maoka, T., Takemura, M., Tokuda, H., Suzuki, N., & Misawa, N. (2014). 4-Ketozeinoxanthin, a novel carotenoid produced in Escherichia coli through metabolic engineering using carotenogenic genes of bacterium and liverwort. *Tetrahedron Letters, 55*(40), 6708–6710.

Marienhagen, J., & Bott, M. (2013). Metabolic engineering of microorganisms for the synthesis of plant natural products. *Journal of Biotechnology, 163*, 166–178.

Marín, L., Gutiérrez-del-Río, I., Entrialgo-Cadierno, R., Villar, C. J., & Lombó, F. (2018). De novo biosynthesis of myricetin, kaempferol, and quercetin in Streptomyces albus and Streptomyces coelicolor. *PLoS One, 13*(11), e0207278. https://doi.org/10.1371/journal.pone.0207278

Mata-Gómez, L. C., Montañez, J. C., Méndez-Zavala, A., & Aguilar, C. N. (2014). Biotechnological production of carotenoids by yeasts: An overview. *Microbial Cell Factories, 13*, 12. https://doi.org/10.1186/1475-2859-13-12

McWilliams, A. (2018). *The global market for carotenoids*. Retrieved from https://www.bccresearch.com/market-research/food-and-beverage/the-global-market-for-carotenoids-fod025f.html

Meng, H., Wang, H., & Xiong, Z. (2020). Stable and marker-free genetic engineering of bacteria using an auto-selection marker system. *Frontiers in Bioengineering and Biotechnology, 8*, 578.

Mikkelsen, M. D., Buron, L. D., Salomonsen, B., Olsen, C. E., Hansen, B. G., Mortensen, U. H., & Halkier, B. A. (2012). Microbial production of indolylglucosinolate through engineering of a multi-gene pathway in a versatile yeast expression platform. *Metabolic Engineering, 14*, 104–111.

Miyahisa, I., Funa, N., Ohnishi, Y., Martens, S., Moriguchi, T., & Horinouchi, S. (2006). Combinatorial biosynthesis of flavones and flavonols in Escherichia coli. *Applied Microbiology and Biotechnology, 71*, 53–58.

Mohammed, A. S. A., Naveed, M., & Jost, N. (2021). Polysaccharides; classification, chemical properties, and future perspective applications in fields of pharmacology and biological medicine (a review of current applications and upcoming potentialities). *Journal of Polymers and the Environment, 29*, 2359–2371. https://doi.org/10.1007/s10924-021-02052-2

Mohd Nadzir, M., Nurhayati, R. W., Idris, F. N., & Nguyen, M. H. (2021). Biomedical applications of bacterial exopolysaccharides: A review. *Polymers, 13*(4), 530. https://doi.org/10.3390/polym13040530

Mora-Pale, M., Sanchez-Rodriguez, S. P., Linhardt, R. J., Dordick, J. S., & Koffas, M. A. (2013). Metabolic engineering and in vitro biosynthesis of phytochemicals and non-natural analogues. *Plant Science, 210*, 10–24.

Nakagawa, A., Matsuzaki, C., Matsumura, E., Koyanagi, T., Katayama, T., Yamamoto, K., Sato, F., Kumagai, H., & Minami, H. (2014). (R, S)-Tetrahydropapaveroline production by stepwise fermentation using engineered Escherichia coli. *Scientific Reports, 4*.

Nakagawa, A., Minami, H., Kim, J.-S., Koyanagi, T., Katayama, T., Sato, F., & Kumagai, H. (2011). A bacterial platform for fermentative production of plant alkaloids. *Nature Communications, 2*, 326.

Nakata, R., Takahashi, S., & Inoue, H. (2012). Recent advances in the study on resveratrol. *Biological and Pharmaceutical Bulletin, 35*, 273–279.

Netzer, R., Stafsnes, M. H., Andreassen, T., Goksøyr, A., Bruheim, P., & Brautaset, T. (2010). Biosynthetic pathway for γ-cyclic sarcinaxanthin in Micrococcus luteus: Heterologous expression and evidence for diverse and multiple catalytic functions of C50 carotenoid cyclases. *Journal of Bacteriology, 192*, 5688–5699.

Nielsen, J. C., & Nielsen, J. (2017). Development of fungal cell factories for the production of secondary metabolites: Linking genomics and metabolism. *Synthetic and Systems Biotechnology, 2*, 5–12. https://doi.org/10.1016/j.synbio.2017.02.002

Orikasa, Y., Tanaka, M., & Sugihara, S. et al. (2009). pfaB products determine the molecular species produced in bacterial polyunsaturated fatty acid biosynthesis. *FEMS Microbiology Letters, 295*, 170–176. https://doi.org/10.1111/j.1574-6968.2009.01582.x

Palmer, C. M., & Alper, H. S. (2018). Expanding the chemical palette of industrial microbes: Metabolic engineering for type III pks-derived polyketides. *Biotechnology Journal, 14*, e1700463.

Panche, A., Diwan, A., & Chandra, S. (2016). Flavonoids: An overview. *Journal of Nutritional Science, 5*, e47. https://doi.org/10.1017/jns.2016.41

Pandey, R. P., Parajuli, P., Koffas, M. A. G., & Sohng, J. K. (2016). Microbial production of natural and non-natural flavonoids: Pathway engineering, directed evolution and systems/synthetic biology. *Biotechnology Advances, 34*, 634–662.

Park, S. R., Yoon, J. A., Paik, J. H., Park, J. W., Jung, W. S., Ban, Y. H., Kim, E. J., Yoo, Y. J., Han, A. R., & Yoon, Y. J. (2009). Engineering of plant-specific phenylpropanoids biosynthesis in Streptomyces venezuelae. *Journal of Biotechnology, 141*(3–4), 181–188.

Parra Daza, L. E., Suarez Medina, L., Tafur Rangel, A. E., Fernández-Niño, M., Mejía-Manzano, L. A., González-Valdez, J., Reyes, L. H., & González Barrios, A. F. (2023). Design and assembly of a biofactory for (2S)-Naringenin production in Escherichia coli: Effects of oxygen transfer on yield and gene expression. *Biomolecules, 13*(3), 565.

Peng, Y.-F., Chen, W.-C., Xiao, K., Xu, L., Wang, L., & Wan, X. (2016). DHA production in Escherichia coli by expressing reconstituted key genes of polyketide synthase pathway from marine bacteria. *PLoS One, 11*, e0162861.

Phelan, R. M., Sekurova, O. N., Keasling, J. D., & Zotchev, S. B. (2015). Engineering terpene biosynthesis in Streptomyces for production of the advanced biofuel precursor bisabolene. *ACS Synthetic Biology, 4*(4), 393–399.

Pieters, B. J., van den Berg, W. A., Heringa, J., & Krab, K. (2021). Multiplexed gene expression using low-cost antibiotics-free vectors and fast microfluidic-based screen-and-sort assays. *Biotechnology and Bioengineering, 118*(3), 805–817.

Pineiro, M., Asp, N.-G., Reid, G., Macfarlane, S., Morelli, L., Brunser, O., & Tuohy, K. (2008). FAO technical meeting on prebiotics. *Journal of Clinical Gastroenterology, 42*, S156–S159. https://doi.org/10.1097/MCG.0b013e31815a064e

Rajkumar, A. S., Liu, G., Bergenholm, D., Arsovska, D., Kristensen, M., Nielsen, J., Jensen, M. K., Keasling, J. D., & Borodina, I. (2016). Engineering of synthetic, stress-responsive yeast promoters. *Nucleic Acids Research, 44*(13), e136.

Rao, A. V., & Rao, L. G. (2007). Carotenoids and human health. *Pharmacological Research, 55*(3), 207–216. https://doi.org/10.1016/j.phrs.2007.01.012

Rinkel, J., Lauterbach, L., & Dickschat, J. S. (2017). Spata-13,17-diene synthase—An enzyme with sesqui-, di-, and sesterterpene synthase activity from Streptomyces xinghaiensis. *Angewandte Chemie International Edition, 56*(52), 16385–16389.

Rodriguez, A., Strucko, T., Stahlhut, S. G., Kristensen, M., Svenssen, D. K., Forster, J., Nielsen, J., & Borodina, I. (2017). Metabolic engineering of yeast for fermentative production of flavonoids. *Bioresource Technology, 245*, 1645–1654.

Rodriguez-Colinas, B., de Abreu, M. A., Fernandez-Arrojo, L., de Beer, R., Poveda, A., Jimenez-Barbero, J., Haltrich, D., Ballesteros Olmo, A. O., Fernandez-Lobato, M., & Plou, F. J. (2011). Production of galacto-oligosaccharides by the β-galactosidase from Kluyveromyces lactis: Comparative analysis of permeabilized cells versus soluble enzyme. *Journal of Agricultural and Food Chemistry, 59*, 10477–10484. https://doi.org/10.1021/jf2029629

Russo, G. L., Russo, M., Spagnuolo, C., Tedesco, I., Bilotto, S., Iannitti, R., & Palumbo, R. (2014). Quercetin: A pleiotropic kinase inhibitor against cancer. *Cancer Treatment and Research, 159*, 185–205. https://doi.org/10.1007/978-3-642-38007-5_11

Sankari, M., Rao, P. R., Hemachandran, H., Pullela, P. K., Doss, C. G. P., Tayubi, I. A., et al. (2018). Prospects and progress in the production of valuable carotenoids: Insights from metabolic engineering, synthetic biology, and computational approaches. *Journal of Biotechnology, 266*, 89–101. https://doi.org/10.1016/j.jbiotec.2017.12.010

Sathish, T., & Prakasham, R. S. (2013). Intensification of fructosyltransferases and fructo-oligosaccharides production in solid-state fermentation by aspergillus awamori GHRTS. *Indian Journal of Microbiology, 53*(3), 337–342. https://doi.org/10.1007/s12088-013-0380-5

Segall-Shapiro, T. H., Sontag, E. D., & Voigt, C. A. (2018). Engineered promoters enable constant gene expression at any copy number in bacteria. *Nature Biotechnology, 36*(4), 352–358.

Shakya, A. K. (2020). *Unit-7 polysaccharides*. Indira Gandhi National Open University.

Shen, Y., Zhao, L., Li, Y. et al. (2014). Synthesis of β-alanine from L-aspartate using L-aspartate-α-decarboxylase from Corynebacterium glutamicum. *Biotechnol Lett 36*, 1681–1686. https://doi.org/10.1007/s10529-014-1527-0

Shi, Y. H., Pan, L. H., Zhang, W. J., Li, H. J., & Wu, J. C. (2007). Optimization of culture conditions for γ-polyglutamic acid production by Bacillus subtilis ZJU-7. *Biotechnology and Applied Biochemistry, 47*(2), 113–118.

Shima, S., & Sakai, H. (1977). Polylysine produced by Streptomyces. *Agricultural and Biological Chemistry, 41*(9), 1907–1909. https://doi.org/10.1271/bbb1961.41.1907

Shin, S. Y., Han, N. S., Park, Y. C., Kim, M. D., & Seo, J. H. (2011). Production of resveratrol from p-coumaric acid in recombinant Saccharomyces cerevisiae expressing 4-coumarate: Coenzyme a ligase and stilbene synthase genes. *Enzyme and Microbial Technology, 48*, 48–53.

Soma, Y., Fujiwara, Y., Nakagawa, T., Tsuruno, K., & Hanai, T. (2017). Reconstruction of a metabolic regulatory network in Escherichia coli for purposeful switching from cell growth mode to production mode in direct GABA fermentation from glucose. *Metabolic Engineering, 43*, 54–63.

Song, C. W., Lee, J., Ko, Y. S., & Lee, S. Y. (2015). Metabolic engineering of Escherichia coli for the production of 3-aminopropionic acid. *Metabolic Engineering, 30*, 121–129.

Soumya, M. P., & Nampoothiri, K. M. (2021). An overview of functional genomics and relevance of glycosyltransferases in exopolysaccharide production by lactic acid bacteria. *International Journal of Biological Macromolecules, 184*, 1014–1025. https://doi.org/10.1016/j.ijbiomac.2021.06.131

Sundstrom, E., Yaegashi, J., Yan, J., Masson, F., Papa, G., Rodriguez, A., Bhan, N., Sale, K., Smith, D., Brem, R., et al. (2018). Demonstrating a separation-free process coupling ionic liquid pretreatment, saccharification, and fermentation with Rhodosporidium toruloides to produce advanced biofuels. *Green Chemistry, 20*(12), 2870–2879.

Survase, S. A., Saudagar, P. S., & Singhal, R. S. (2007). Enhanced production of scleroglucan by Sclerotium rolfsii MTCC 2156 by use of metabolic precursors. *Bioresource Technology, 98*, 410–415.

Tian, F., Khodadadi, M., & Karboune, S. (2014). Optimization of levansucrase/endo-inulinase bi-enzymatic system for the production of fructooligosaccharides and oligolevans from sucrose. *Journal of Molecular Catalysis B: Enzymatic, 109*, 85–93. https://doi.org/10.1016/j.molcatb.2014.08.005

Trantas, E., Panopoulos, N., & Ververidis, F. (2009). Metabolic engineering of the complete pathway leading to heterologous biosynthesis of various flavonoids and stilbenoids in Saccharomyces cerevisiae. *Metabolic Engineering, 11*(6), 355–366. https://doi.org/10.1016/j.ymben.2009.07.004

Tsao, C. Y., Hooshangi, S., Wu, H. C., Valdes, J. J., & Bentley, W. E. (2010). Autonomous induction of recombinant proteins by minimally rewiring native quorum sensing regulon of E. coli. *Metabolic Engineering, 12*(3), 291–297.

Tseng, C. H., Yang, C. Y., & Lin, S. P. (2012a). Overexpression of cyanophycin synthetase in Escherichia coli for enhanced cyanophycin production. *Journal of Industrial Microbiology and Biotechnology, 39*(3), 449–456.

Tseng, W. C., Fang, T. Y., Cho, C. Y., Chen, P. S., & Tsai, C. S. (2012b). Assessments of growth conditions on the production of cyanophycin by recombinant Escherichia coli strains expressing cyanophycin synthetase gene. *Biotechnology Progress, 28*, 358–363.

Vaishnav, A. M., Upadhyay, K. H., Tipre, D. R., & Dave, S. R. (2021). Bio-prospecting of fruit waste for exopolysaccharide production by bacteria. In *Biotechnology for sustainable environment* (pp. 353–371). Springer.

Van Summeren-Wesenhagen, P. V., & Marienhagen, J. (2015). Metabolic engineering of Escherichia coli for the synthesis of the plant polyphenol pinosylvin. *Applied and Environmental Microbiology, 81*, 840–849.

Vera, C., Guerrero, C., Illanes, A., & Conejeros, R. (2014). Fed-batch synthesis of galacto-oligosaccharides with aspergillus oryzae β-galactosidase using optimal control strategy. *Biotechnology Progress, 30*(1), 59–67. https://doi.org/10.1002/btpr.1831

Wang, P., Ma, J., Zhang, Y., Zhang, M., Wu, M., Dai, Z., & Jiang, M. (2016). Efficient secretory overexpression of endoinulinase in Escherichia coli and the production of inulooligosaccharides. *Applied Biochemistry and Biotechnology, 179*(5), 880–894. https://doi.org/10.1007/s12010-016-2037-4

Wang, S., Zhang, S., Xiao, A., Rasmussen, M., Skidmore, C., & Zhan, J. (2015). Metabolic engineering of Escherichia coli for the biosynthesis of various phenylpropanoid derivatives. *Metabolic Engineering, 29*, 153–159.

Wang, Y. Y., Xu, J. Z., & Zhang, W. G. (2019a). Metabolic engineering of l-leucine production in Escherichia coli and Corynebacterium glutamicum: A review. *Critical Reviews in Biotechnology, 39*(6), 781–795.

Wang, Y. Y., Zhang, F., Xu, J. Z., Zhang, W. G., Chen, X. L., & Liu, L. M. (2019b). Improvement of l-leucine production in Corynebacterium glutamicum by altering the redox flux. *International Journal of Molecular Sciences, 20*(8), 2020.

Wiseman, B., Nitharwal, R. G., Widmalm, G., & Högbom, M. (2021). Structure of a full-length bacterial polysaccharide co-polymerase. *Nature Communications, 12*(1), 369. https://doi.org/10.1038/s41467-020-20579-1

Wu, J., Du, G., Zhou, J., & Chen, J. (2013a). Metabolic engineering of Escherichia coli for (2S)-pinocembrin production from glucose by a modular metabolic strategy. *Metabolic Engineering, 16*, 48–55.

Wu, J., Liu, P., Fan, Y., Bao, H., Du, G., Zhou, J., & Chen, J. (2013b). Multivariate modular metabolic engineering of Escherichia coli to produce resveratrol from L-tyrosine. *Journal of Biotechnology, 167*, 404–411.

Wu, J., Zhou, T., Du, G., Zhou, J., & Chen, J. (2014). Modular optimization of heterologous pathways for de novo synthesis of (2S)-naringenin in Escherichia coli. *PLoS One, 9*, e101492.

Wu, Q., & Shah, N. P. (2017). High γ-aminobutyric acid production from lactic acid bacteria: Emphasis on Lactobacillus brevis as a functional dairy starter. *Critical Reviews in Food Science and Nutrition, 57*, 3661–3672.

Xie, D., Jackson, E. N., & Zhu, Q. (2015). Sustainable source of omega-3 eicosapentaenoic acid from metabolically engineered Yarrowia lipolytica: From fundamental research to commercial production. *Applied Microbiology and Biotechnology, 99*, 1599–1610.

Xu, P., Ranganathan, S., Fowler, Z. L., Maranas, C. D., & Koffas, M. A. (2011). Genome-scale metabolic network modeling results in minimal interventions that cooperatively force carbon flux towards malonyl-CoA. *Metabolic Engineering, 13*, 578–587.

Xue, Z., Sharpe, P. L., Hong, S.-P., et al. (2013). Production of omega-3 eicosapentaenoic acid by metabolic engineering of Yarrowia lipolytica. *Nature Biotechnology, 31*, 734–740. https://doi.org/10.1038/nbt.2622

Yaegashi, J., Kirby, J., Ito, M., Sun, J., Dutta, T., Mirsiaghi, M., Pray, T. R., Chen, W., Li, J., Meichenin, A., et al. (2017). Rhodosporidium toruloides: A new platform organism for conversion of lignocellulose into terpene biofuels and bioproducts. *Biotechnology for Biofuels, 10*, 241.

Yang, J., & Guo, L. (2014). Biosynthesis of β-carotene in engineered *E. coli* using the MEP and MVA pathways. *Microbial Cell Factories, 13*, 160. https://doi.org/10.1186/s12934-014-0160-5

Yang, Y., Lin, Y., Li, L., Linhardt, R. J., & Yan, Y. (2015). Regulating malonyl-CoA metabolism via synthetic antisense RNAs for enhanced biosynthesis of natural products. *Metabolic Engineering, 29*, 217–226.

Yoshimura, T., Shibata, N., Hamano, Y., & Yamanaka, K. (2015). Heterologous production of hyaluronic acid in an ε-poly-L-lysine producer, Streptomyces albulus. *Applied and Environmental Microbiology, 81*(10), 3631–3640. https://doi.org/10.1128/AEM.03807-14

Yu, H., & Stephanopoulos, G. (2008). Metabolic engineering of Escherichia coli for biosynthesis of hyaluronic acid. *Metabolic Engineering, 10*(1), 24–32. https://doi.org/10.1016/j.ymben.2007.10.003

Yu, L., & O'Sullivan, D. J. (2014). Production of galactooligosaccharides using a hyperthermophilic β-galactosidase in permeabilized whole cells of Lactococcus lactis. *Journal of Dairy Science, 97*(2), 694–703. https://doi.org/10.3168/jds.2013-7506

Zargar, A., Quan, D. N., Emamian, M., Tsao, C. Y., Wu, H. C., Virgile, C. R., & Bentley, W. E. (2015). Rational design of 'controller cells' to manipulate protein and phenotype expression. *Metabolic Engineering, 30*, 61–68.

Zhang, C., Liu, L., Teng, L., Chen, J., Liu, J., Li, J., Du, G., & Chen, J. (2012). Metabolic engineering of Escherichia coli BL21 for biosynthesis of heparosan, a bioengineered heparin precursor. *Metabolic Engineering, 14*(5), 521–527. https://doi.org/10.1016/j.ymben.2012.06.004

Zhang, F., Liu, J., & Shi, J.-S. (2010). Anti-inflammatory activities of resveratrol in the brain: Role of resveratrol in microglial activation. *European Journal of Pharmacology, 636*, 1–7.

Zhang, W., Liu, H., Li, X., Liu, D., Dong, X.-T., Li, F.-F., Wang, E.-X., Li, B.-Z., & Yuan, Y.-J. (2017). Production of naringenin from d-xylose with co-culture of *E. coli* and S. cerevisiae. *Engineering in Life Sciences, 17*, 1021–1029.

Zhang, W., Yang, Y., Liu, X., Liu, C., & Bai, Z. (2019). Development of a secretory expression system with high compatibility between expression elements and an optimized host for endoxylanase production in Corynebacterium glutamicum. *Microbial Cell Factories, 18*, 1–13.

Zhang, Y., Li, S. Z., Li, J., Pan, X., Cahoon, R. E., Jaworski, J. G., Wang, X., Jez, J. M., Chen, F., & Yu, O. (2006). Using unnatural protein fusions to engineer resveratrol biosynthesis in yeast and mammalian cells. *Journal of the American Chemical Society, 128*, 13030–13031.

Zhao, J., Li, Q., Sun, T., Zhu, X., Xu, H., Tang, J., Zhang, X., & Ma, Y. (2013). Engineering central metabolic modules of Escherichia coli for improving β-carotene production. *Metabolic Engineering, 17*, 42–50.

Zhao, S., Jones, J. A., Lachance, D. M., Bhan, N., Khalidi, O., Venkataraman, S., Wang, Z., & Koffas, M. A. (2015). Improvement of catechin production in Escherichia coli through combinatorial metabolic engineering. *Metabolic Engineering, 28*, 43–53.

Zhou, S., Hao, T., & Zhou, J. (2020). Fermentation and metabolic pathway optimization to de novo synthesize (2S)-Naringenin in Escherichia coli. *Journal of Microbiology and Biotechnology, 30*(10), 1574–1582.

Zhu, F., Lu, L., Fu, S., Zhong, X., Hu, M., Deng, Z., & Liu, T. (2015). Targeted engineering and scale up of lycopene overproduction in Escherichia coli. *Process Biochemistry, 50*, 341–346. https://doi.org/10.1016/j.procbio.2014.12.012

Zhu, Q., & Jackson, E. N. (2015). Metabolic engineering of Yarrowia lipolytica for industrial applications. *Current Opinion in Biotechnology, 36*, 65–72.

Zhu, S., Wu, J., Du, G., Zhou, J., & Chen, J. (2014). Efficient synthesis of eriodictyol from L-tyrosine in *Escherichia coli. Applied and Environmental Microbiology, 80*, 3072–3080.

Zou, X., Guo, L., Huang, L., Li, M., Zhang, S., Yang, A., Zhang, Y., Zhu, L., Zhang, H., Zhang, J., & Feng, Z. (2020). Pathway construction and metabolic engineering for fermentative production of beta-alanine in Escherichia coli. *Applied Microbiology and Biotechnology, 104*(6):2545–59.

# Chapter 2
# Seaweeds as Potential Source of Bioactive Compounds

**Contents**



## 2.1 Introduction

A plentiful supply of structurally unique and physiologically active metabolites can be found in marine species. Any big marine benthic algae that is multicellular, macrothallic, and macroscopic that is attached to the ocean floor or solid substrate is referred to as seaweed. This distinguishes seaweeds from the majority of algae, which are tiny (Leelavathi & Prasad, 2014). Algae, including seaweeds (marine macroalgae) and microalgae (phytoplankton), contribute nearly 30% of world aquaculture production (measured in wet weight), primarily from seaweeds. Seaweeds are also called macroalgae. In contrast, microalgae (Cyanophyceae) are small and unicellular aquatic photosynthetic plant-like organisms. Seaweeds and microalgae generate socio-economic benefits to tens of thousands of households, primarily in coastal communities, including numerous women empowered by seaweed cultivation. An important part of the cuisine in many Asian nations is seaweed. It has been demonstrated that one of the factors influencing Japan and China's lower incidence of breast and prostate cancer than that of North America and Europe is their diet's high seaweed consumption. Since seaweeds are frequently utilised as vegetables,




P. Saini, P. Mishra, *Bioprospecting of Microbial Resources for Agriculture, Environment and Bio-chemical Industry*, https://doi.org/10.1007/978-3-031-63844-2_2

spices, wrappers for sushi, and condiments, most Japanese people eat between 10% and 25% of their diets made up of seaweed (Leelavathi & Prasad, 2014). Seaweeds are low-value goods that are not well known in many other parts of the world because they are mostly grown in Eastern and Southeast Asia. China harvests over 5 million wet tonnes of edible seaweeds, making it the world's largest producer. The majority of this goes towards kombu, which is made from hundreds of hectares of brown seaweed called Laminaria japonica that is hung on ropes in the water. About half of the 800,000 wet tones of three different species that are farmed in the Republic of Korea are used to manufacture wakame, which is made from *Undaria pinnatifida*, a separate type of brown seaweed that is grown similarly to *Laminaria* in China. Of the approximately 600,000 wet tonnes produced in Japan, 75% is used to make nori, the thin, black seaweed used to wrap sushi rice balls (FAO, 2021). For a range of applications, including culinary, cosmetic, pharmaceutical, and biotechnological, but with a small market base as in the case of novel meals, micronutrient supplements, and plant-based diets for human and animal welfare. Seaweeds are considered a potential functional food that may reduce the risk of many diseases due to their diverse range of bioactive chemicals. Most seaweed species are edible and don't have any inherent poisons. Around 700 edible seaweed species, including roughly 195 brown seaweeds, 345 red seaweeds, and 125 green seaweeds, have been described. Seaweed usage by humans extends back hundreds of years (Pereira, 2016). Earlier few numbers of species and chemicals are often the focus of scientific studies on the bioactive substances found in seaweed (Holdt & Kraan, 2011). But for now, seaweeds are thought to be a rich source of structurally distinct bioactive metabolites for the development of novel functional food-based pharmacophores and medicines, and their shown health advantages have drawn a lot of interest to them.

People outside of Eastern Asia typically have little or no exposure to or preference over consuming seaweed, even though the majority of seaweed produced there is directly consumed as food. Demand for seaweed would need to rise to support seaweed production expansion. Increasing the amount of seaweed consumed as food for humans is one strategy, as this tends to make better use of seaweed and boost revenue for seaweed growers. The adaptability and diversity of seaweeds indicate that they can be used in a wide range of food products, giving food producers and distributors more options for nutrient-dense, low-calorie, and healthful foods. Therefore, the purpose of this chapter is to present a comprehensive overview of the seaweed's importance, application, and main biochemicals that have been isolated from different species of seaweed together with their biological characteristics. These substances were discovered to have a wide range of pharmacological characteristics and inhibitory enzyme activity against important enzymes implicated in the genesis of non-communicable diseases. Future research should focus more on their toxicity, clinical efficacy, mechanisms of action, and interactions with traditional foods, as these areas are still poorly studied (Rengasamy et al., 2020).

## 2.2 Classification and Characteristic Features of Seaweeds

Any macroscopic, multicellular, or benthic marine algae are referred to as seaweed. Seaweeds are a kind of photosynthetic, taxonomically varied marine plants (non-flowering plant-like organisms also known as macroalgae,) that inhabit the sea and from which terrestrial plants separated about 50 billion years ago. Based on pigmentation, seaweeds can be divided into three major categories: brown, red, and green. These three major groupings are referred to by botanists as Phaeophyceae, Rhodophyceae, and Chlorophyceae, in that order. Seaweed may be categorized into one of multi cell algae and according to the presence of pigment in it, they may be Chlorophyceae (green algae), Pheophyceae (brown algae), and Rhodophyceae (red algae) (Percival, 1979). Like land plants, green algae need plenty of light and contain chlorophylla and b. As a result, littoral seas are home to the majority of these taxa. Chlorophyll a and d are present in red seaweeds and phycobilins, which include phycocyanin and phycoerythrin, which together give thallus its distinctive red hue. Red seaweeds, in contrast to green seaweeds, can survive in extremely low light conditions and are therefore found in the lowest underwater regions where only blue light can reach. Fucoxanthin and chlorophyll a and c are present in brown algae, giving it its brown color. It needs sunshine much like green seaweed, but not as much of it.

Seaweeds are different from other marine plants like mangroves and seagrasses since they don't have actual vascular tissue, leaves, stems, roots, or flowers. The basic structure of seaweed depicts the foliose blade, also known as the frond, the stem-like stipe, and the holdfast that affixes the plant to a hard substrate makeup. Thus, the term "thallus" refers to the entire plant. A significant amount of form diversity is provided by variations in the thallus' structure. The thallus morphology of seaweeds varies greatly, encompassing basic crusts, leafy blades, tubular to filamentous forms with simple branching, as well as more sophisticated forms. Seaweeds vary in size from several centimeters to several meters. The enormous kelps in temperate regions can reach a height of 60 meters. Seaweeds get their nourishment from containing additional substances and micronutrients straight from the sea to support their development, growth, and proliferation. As a result, the growth of these plants typically depends on the availability of sunshine and nutrients; the growth and total productivity of seaweeds would be restricted in the absence of any of these.

Identification of a species is crucial before researching the activity of seaweed's bioactive components. Seaweeds were identified by botanical methods to analyze their morphological characteristics as a preliminary technique (Arif et al., 2021). Molecular identification technology is more systematic and reliable which nowadays is employed to identify seaweeds. Consequently, the application of molecular biology techniques is necessary to reveal putative species, which is advantageous for the ensuing studies about the bioactive components of seaweeds (Yang et al., 2021).

## 2.3 Bioactive Compounds Derived from Seaweed

According to Hentati et al. (2020), the marine ecosystem is thought to be a great source of natural chemicals with a variety of potential uses. According to Kolanjinathan et al. (2014), seaweeds are marine plant organisms that can produce a broad variety of active metabolites with a wide range of medical applications. They also use these metabolites to protect themselves against other invading species. Although seaweeds are well-known for being rich in minerals, vitamins, and polysaccharides (Arasaki & Arasaki, 1983; Rengasamy et al. 2020), they also contain bioactive substances such as proteins, lipids, polyphenols, and polysaccharides that have numerous other properties, including antifungal, antiviral, and antibacterial qualities (Kumar et al., 2008). Sulfuric acid polysaccharides, sulphated galactans, and xylans are found in green algae. Sargassan, laminarin (β-1,3 glucan), fucoidan (sulphated fucose), and alginic acid are all found in brown algae. Agars, carrageenans, xylans, floridean starch (glucan that resembles amylopectin), water-soluble sulphated galactan, and porphyran are among the mucopolysaccharides found in the intercellular spaces of red algae.Given the significance of seaweeds, several environmentally friendly extraction techniques, including Pressurized Solvents Extraction (PSE), Enzyme-assisted Extraction (EAE), Supercritical Fluid Extraction (SFE), Microwave-assisted Extraction (MAE), Ultrasound-Assisted Extraction (UAE) and Enzyme-assisted Extraction (EAE), were employed in place of the traditional method to isolate the bioactive components. The extract was then further purified using chromatographic technique analysis to guarantee its purity. (Amlani & Yetgin, 2022). Numerous biotechnological applications make use of the bioactive chemicals found in seaweeds (Lomartire et al., 2021). Applications involving the bioactive chemicals found in seaweeds for pharmacological, biological, and nutraceutical purposes are further demonstrated (Table 2.1). Owing to these compounds' characteristics, biomedicine and contemporary pharmacy can benefit from their advancements to create novel formulations using ingredients found in nature.

## 2.4 Potential Benefits

The production and consumption/usage of goods based on seaweed will increase with the incorporation of seaweed extracts in the food, nutraceutical, pharmaceutical, and industrial sectors. "Nutraceutical food" is a term used to describe food that has potential health benefits in addition to its nutritional value. Foods containing seaweed are regarded as nutraceuticals because of their beneficial benefits on human health, which include the relief of autoimmune, cardiovascular, ophthalmic, and diabetic disorders as well as arthritis. Figure 2.1 provides a schematic representation of the extraction of bioactive compounds and its potential health benefits. Phlorotannins derived from *E. cava* stimulate the GABAA–BZD receptor to induce

**Table 2.1** Potential application of bioactive compounds derived from seaweeds and its health benefits

| S. no. | Bioactive compound | Seaweed | Health benefits | Application | References |
|---|---|---|---|---|---|
| 1. | Phycocolloids | brown and red algae | antitumoral activity | gelling agents in food, dressings, coatings of medicaments, culture medium, cosmetics | Percival (1979) |
| 2. | Fucoidans (Sulphated polysaccharides), Phlorotannin (Polyphenolic compound), Fucoxanthin, Alginate, Alginic acid, Laminaran, linoleic acid (C18:2n-6, LA), arachidonic acid (C20:4n-6, DHA) and eicosapentaenoic acid (C22:5n-3, EPA) | Phaeophyceae (brown algae) *Laminaria spp* and *Ecklonia spp, Saccharina longicruris, Ascophyllum nodosum and Fucus vesiculosus, Durvillaea spp., Lessonia spp., Macrocystis spp.,* and *Sargassum spp.* | anti-inflammatory, anticoagulant, and antithrombotic effects, anti-obesity properties, biodegradable, biocompatible, non-toxic | Nutraceutical and cosmetic | Cumashi et al. (2007) |
| 3. | Ulvan (a sulphate polysaccharide), Palmitic acid (saturated fatty acid), Linoleic acid (unsaturated fatty acid), Pigments like Chlorophylls (a and b) and Carotenoids (β-carotene and xanthophylls) | Chlorophyta (green algae) *Codium* spp., *Ulva* spp. and *Chaetomorpha* spp | natural drugs, antioxidant activity | biomedicine, cosmetic, agricultural and pharmaceuticalemulsifiers, stabilizers, and thickeners in food productsmedical applications and production of alternative biomaterial | Bajpai et al. (2014) |
| 4. | carrageenans and agar, sesquiterpenes, diterpenes, triterpenes, desmosterol, cholesterol, sitosterol, fucosterol and chalinastero | Rhodophyta (red algae)*Chondrus pinnulatus, Chondrus armatus, Chondrus yendoi, Mastocarpus pacificus, Mazzaella hemisphaerica* | anticoagulant agents, antitumoral therapies, antitumor immunity activation, antifungal, and antibacterial activity | medical and pharmaceutical sectors | Pal et al. (2014) |

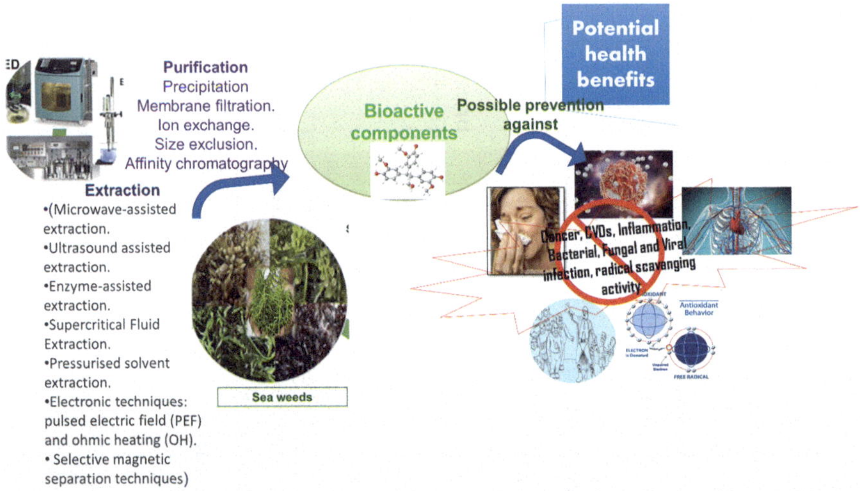


**Fig. 2.1** Schematic representation of extraction of bioactive compounds and its potential health benefits

sleep. The GABAA–BZD receptor is a well-known molecule for sedatives and hypnotics (Cho et al., 2012). Seaweeds develop at a significantly faster rate than other food sources and plants. Although they can be found naturally, seaweeds are being farmed more and more these days to exploit their antioxidant properties. Seaweed is a major source of antioxidant chemicals, according to several research findings (Tanna et al., 2020). Certain natural pigments found in seaweeds, such as chlorophylls, phycobiliproteins, andcarotenoids. Thecarotenoidscan react quickly to reactive oxygen species that act as free radicals, reducing the amount of oxidative degradation and increasing the nutritional value of food. Recent research has shown the great nutritional worth of seaweeds as well as the potent qualities that their bioactive components offer. Species belonging to the class Phaeophyceae, phylum Rhodophyta, and phylum Chlorophyta have special chemicals with several qualities that could be beneficial to human health, making them valuable compounds to use in biotechnological applications (Lomartire et al., 2021). Three major hydrocolloids are derived from different types of red and brown seaweeds: carrageenan, alginate, and agar. A hydrocolloid is a non-crystalline material that dissolves in water to produce a viscous, thicker solution. Its molecules are quite big. Water-soluble carbohydrates like carrageenan, agar, and alginate are used to create gels (jellies) with different levels of firmness, thicken (increase the viscosity of) aqueous solutions, and stabilize certain products like ice cream (they prevent the formation of large ice crystals so that the ice cream can maintain a smooth texture). Seaweeds are processed into food additives or supplements in addition to being consumed directly by humans (McHugh, 2003; FAO, 2018). One of the additives monosodium glutamate, a flavour enhancer that is frequently used in products, was derived from Japanese kelp (*Laminaria japonica*) to attainthe taste of umami (Milinovic et al., 2021). Seaweed-based hydrocolloids are widely used as food additives to improve the quality of a variety of foods. Agar is extracted from *Gracilaria* and other *agarophytes*, carrageenan is extracted from *Kappaphycus/Eucheuma* and other carrageenophytes, and alginate is extracted from *Laminaria/Saccharina* and other brown seaweeds. Furthermore, food supplements containing seaweed extracts such as iodine, fucoidan, fucoxanthin, and phlorotannins are utilized to promote health. Seaweed has been used as fertilizer at least since the nineteenth century. Coastal residents used it initially, gathering storm-cast seaweed, which was often huge brown seaweed, and digging it into local soil. The seaweed's high fiber content helps retain moisture and conditions the soil, while its mineral content serves as a helpful fertilizer and source of trace elements. Cosmetic items, such as lotions, creams, and toothpaste, occasionally list "marine extract," "extract of alga," "seaweed extract," or something along those lines as ingredients. Seaweeds are regarded as a nutrient-rich food because they are high in minerals, water-soluble (C, B1, B2, B9, B12), and fat-soluble vitamins (A, D, E, and K), essential minerals (calcium, iron, iodine, magnesium, phosphorus, potassium, zinc, copper, manganese, selenium, and fluoride), low in fat content enriched in polyunsaturated fatty acids, protein, essential amino acids, and polyphenols that have anti-inflammatory and antioxidant properties (Lomartire et al., 2021). The iodine level of seaweed can vary depending on its kind and growing environment, making it a potentially unstable source. Additionally, seaweeds are a resource that is underutilized that the food industry might investigate for use as condiments, additions, or supplemental ingredients to enhance the flavour of processed goods.

Some of the potential health benefits are listed below

- Smooth functioning of the thyroid
- Reduced diabetes risk factors
- Support and improve gut health
- Reduced risk of cardiovascular diseases and associated factors
- Helps in weight loss

Other applications and usage of seaweeds are as below

- Fertilizers and soil conditioners
- Animal feed and Fish feed
- Biomass for fuel production
- Cosmetics industry
- Sustainable and Integrated aquaculture
- Wastewater treatment (reduction of nitrogen- and phosphorus-containing compounds and toxic metals removal from industrial wastewater)

## 2.5 Conclusion

Sustainable health will be supported by the industrial and biotechnological applications of seaweed's bioactive components. Therefore, it is critical to continue and expand research on the bioactive chemicals found in seaweeds to identify and mitigate the effects of potentially hazardous compounds and expand the use of seaweed-derived compounds. With the added benefit of exhibiting anticancer, antiviral, antifungal, antidiabetic, antihypertensive, immunomodulatory, anticoagulant, anti-inflammatory, antioxidant, UV-protective, and neuroprotective properties after assimilation, marine seaweeds are an excellent source of bioactive components that support a healthy diet. Regretfully, not all seaweeds meet the requirements to be regarded as food under safety regulations. Seaweeds tend to retain minerals and heavy metals, which, if ingested in food or medication, could harm the health of humans and animals. The main thing preventing people from eating aquatic foods is food safety concerns (Ahern et al., 2021). Consuming farmed and wild seaweeds as well as microalgae may expose people to various risks related to food safety, such as heavy metals or microcystincontaminations) or run the danger of health problems from consuming too much of a particular element (like iodine) (Bouga & Combet, 2015).Therefore, careful investigation has to be done to explore the utilization of the bioactive substances found in seaweeds for biotechnological, food processing, and other industrial purposes will enhance human welfare by encouraging the use of natural products rather than synthetic ones.

## References


Ahern, M., Thilsted, S. H. & Oenema, S. (2021). *The role of aquatic foods in sustainable healthy diets*. Discussion paper. UN Nutrition. 64 pp. Also available at https://www.unnutrition.org/wp-content/uploads/FINAL-UN-Nutrition-Aquatic-foods-Paper_EN_.pdf.

Arasaki, S. & Arasaki, T. (1983). Low calorie, high nutrition vegetables from the sea. *To help you look and feel better, 1*, 6–196.

Amlani, M. Q., & Yetgin, S. (2022). Seaweeds: Bioactive components and properties, potential risk factors, uses, extraction and purification methods. *Marine Science and Technology Bulletin, 11*(1), 9–31.

Arif, M., Li, Y., El-Dalatony, M. M., Zhang, C., Li, X., & Salama, E.-S. (2021). A complete characterization of microalgal biomass through FTIR/TGA/CHNS analysis: An approach for biofuel generation and nutrients removal renew. *Energy, 163*, 1973–1982.

Bajpai, V. K., Rather, I. A., Lim, J., & Park, Y. H. (2014). Diversity of bioactive polysaccharide originated from marine sources: A review. *Indian Journal of Geo-Marine Sciences, 43*, 1857–1869.

Bouga, M., & Combet, E. (2015). Emergence of seaweed and seaweed-containing foods in the UK: focus on labeling, iodine content, toxicity and nutrition. *Food, 4*(2), 240–253.

Cho, S., Yang, H., Jeon, Y. J., Lee, C. J., Jin, Y. H., Baek, N. I., et al. (2012). Phlorotannins of the edible brown seaweed Ecklonia cava Kjellman induce sleep via positive allosteric modulation of gamma-aminobutyric acid type A– benzodiazepine receptor: A novel neurological activity of seaweed polyphenols. *Food Chemistry, 132*, 1133–1142.

Cumashi, A., Ushakova, N. A., Preobrazhenskaya, M. E., D'Incecco, A., Piccoli, A., Totani, L., Tinari, N., Morozevich, G. E., Berman, A. E., Bilan, M. I., et al. (2007). A comparative study of the anti-inflammatory, anticoagulant, antiangiogenic, and antiadhesive activities of nine different fucoidans from brown seaweeds. *Glycobiology, 17*, 541–552.

FAO. (2018). *The global status of seaweed production, trade and utilization*. Globefish Research Programme, Volume 124. Rome. 120 pp. Licence: CC BY-NC-SA 3.0 IGO.

FAO. (2021). *Food and Agriculture Organization of United States. Global seaweeds and microalgae production, 1950–2019*. Food and Agricultural Organization of the United Nations. Rome, June, 2021. pp. 5. https://www.fao.org/fishery/en/publications/280709

Hentati, F., Tounsi, L., Djomdi, D., Pierre, G., Delattre, C., Ursu, A. V., Fendri, I., Abdelkafi, S., & Michaud, P. (2020). Bioactive polysaccharides from seaweeds. *Molecules, 25*(14), 1–29.

Holdt, S. L., & Kraan, S. (2011). Bioactive compounds in seaweed: functional food applications and legislation. *Journal of Applied Phycology, 23*, 543–597.

Kolanjinathan, K., Ganesh, P., & Saranraj, P. (2014). Pharmacological importance of seaweeds: A review. *World Journal of Fish and Marine Sciences, 6*(1), 1–15.

Kumar, C. S., Ganesan, P., Suresh, P. V., & Bhaskar, N. (2008). Seaweeds as a source of nutritionally beneficial compounds—A review. *Journal of Food Science and Technology, 45*, 1–13.

Leelavathi, M. S., & Prasad, M. P. (2014). Evaluation of antioxidant properties of marine sea weed samples by DPPH method Int. *Journal of Pure & Applied Biosciences, 2*(6), 132–137.

Lomartire, S., Marques, J. C., & Gonçalves, A. M. M. (2021). An overview to the health benefits of seaweeds consumption. *Marine Drugs, 19*(6), 341.

McHugh, D. J. (2003). *A guide to the seaweed industry*. Fisheries Technical Paper No. 441. Rome, FAO. 105 pp.

Milinovic, J., Mata, P., Diniz, M., & Noronha, J. P. (2021). Umami taste in edible seaweeds: The current comprehension and perception. *International Journal of Gastronomy and Food Science, 23*, 100301.

Pal, A., Kamthania, M. C., & Kumar, A. (2014). Bioactive compounds and properties of seaweeds—A review. *Open Access Library Journal, 01*, 1–17.

Percival, E. (1979). The polysaccharides of green, red and brown seaweeds: Their basic structure, biosynthesis and function. *British Phycological Journal, 14*, 103–117.

Pereira, L. (2016). *Edible seaweeds of the world* (p. 463). CRC Press.

Rengasamy, K. R. R., Mahomoodally, M. F., Aumeeruddy, M. Z., Zengin, G., Xiao, J., & Kim, D. H. (2020). Bioactive compounds in seaweeds: An overview of their biological properties and safety. *Food and Chemical Toxicology, 135*, 111013.

Tanna, B., Yadav, S., & Mishra, A. (2020). Anti-proliferative and ROS-inhibitory activities reveal the anticancer potential of Caulerpa species. *Molecular Biology Reports, 47*, 7403–7411.

Yang, Y., Zhang, M., Alalawy, A. I., Almutairi, F. M., Al-Duais, M. A., Wang, J., & Salama, E. (2021). Identification and characterization of marine seaweeds for biocompounds production. *Environmental Technology & Innovation, 24*, 101848.

# Chapter 3
# Bioprospecting of Extremophiles for Industrial Enzymes

## Contents





P. Saini, P. Mishra, *Bioprospecting of Microbial Resources for Agriculture, Environment and Bio-chemical Industry*, https://doi.org/10.1007/978-3-031-63844-2_3

## 3.1 Introduction

Extremophiles encompass a diverse array of organisms that have evolved specialized adaptations to not only survive but thrive in environments characterized by extreme conditions. These conditions span a wide spectrum, including extreme temperatures, both high and low, as well as environments with highly acidic or alkaline pH levels. Additionally, they can withstand intense pressures and even exposure to potentially damaging radiation. These remarkable organisms comprise both prokaryotes and eukaryotes, such as microalgae, fungi, lichens, protists, animals, and plants. It's important to note that while true extremophiles are primarily found among the archaea domain, there are also extremophilic organisms within the bacteria and eukaryotes domains.

Across the spectrum of extremophiles, the microbiomes are associated with various distinct phyla. These include Euryarchaeota and Crenarchaeota within the archaea domain. In the bacterial domain, Firmicutes, Proteobacteria, Actinobacteria, and Deinococcus-Thermus are some of the phyla that host extremophilic organisms. Additionally, within the eukaryotic domain, Basidiomycota and Ascomycota are examples of phyla containing extremophiles (Yadav, 2021). This diverse range of microorganisms showcases their incredible ability to adapt and thrive in environments that would typically be considered inhospitable.

Microorganisms found in diverse and extreme habitats have developed specific adaptations to cope with challenging environmental factors like extreme temperatures, salinity, drought, pH levels, UV radiation, and various chemical stresses (Yadav et al., 2016). These microorganisms, known as extremophiles or stress-adaptive microbes, are able to survive and adapt in such demanding conditions. They fall into different categories based on the specific extreme condition they can endure. These categories encompass psychrophiles and psychrotrophs, which can live in cold environments, thermophiles, which thrive in high-temperature settings, alkaliphiles and acidophiles, which flourish in alkaline or acidic environments respectively, peizophiles that endure high-pressure conditions, and halophiles, which can survive in highly saline environments. Some extremophiles also have capabilities to tolerate metal exposure, radiation, and other stressors (Irwin, 2020). Certain microbial communities, called polyextremophilic microbiomes, can even withstand and thrive in more than one challenging condition simultaneously. The metabolic products produced by these extremophilic microbes possess remarkable qualities, including high catalytic efficiency and stability. They remain functional even in environments with high salinity, alkalinity, low water activity, and oxidative agents. Moreover, these products often exhibit an extended shelf life, making them highly valuable for various industrial applications.

The beneficial extremophilic microbiomes play a crucial role in promoting agro-environmental sustainability. These stress-adaptive microbes possess the ability to produce essential bioactive compounds that are valuable in agricultural and industrial contexts. These compounds include indole acetic acids, gibberellic acids, cytokinins, hydrogen cyanides, ammonia, siderophores, 1-aminocyclopropane-1-carboxylate deaminase, and various extracellular microbial hydrolytic enzymes such as amylase, xylanase, pectinase, and chitinase. These extremophiles find diverse applications across multiple fields including agriculture, dairy, biodegradation, chemical processing, hemicellulose bioconversion, molecular biology, composting, detergents, food, feed additives, leather, paper, cellulose, peptide synthesis, pharmaceuticals, and therapeutic agents (Kumar et al., 2021; Dumorne et al., 2017; Saxena et al., 2020). Their versatility makes them particularly valuable in the food and food processing industry, as they can produce a wide range of bioactive compounds, secondary metabolites, flavors, food ingredients, and vitamins (Barcelos et al., 2020; Kumar et al., 2021).

A notable characteristic of extremophiles is their remarkable capacity to generate an extensive array of valuable products. These encompass enzymes, exopolysaccharides, primary and secondary metabolites, all of which have applications in various industries such as food, detergent, pharmaceuticals, leather, textiles, paper, bioremediation, and agriculture (Mohanty et al., 2022). The biodiversity and distribution of extremophilic microbiomes, including xerophiles, halophiles, acidophiles, thermophiles, alkaliphiles, and psychrophiles, existing in challenging environmental conditions have been extensively investigated over the past decades, with a focus on both culturable and metagenomic approaches (Kumar et al., 2021; Yadav, 2021). Advances in molecular tools like comparative genomics, proteomics, metagenomics, and transcriptomics have revolutionized the study of extremophiles. These tools provide researchers with a deeper insight into the complete metabolic potential of extremophiles, enabling the identification and utilization of their valuable products across a wide range of practical applications (Salwan & Sharma, 2020).

## 3.2 Extremophiles and Their Extreme Habitats

The term "extremophile" refers to organisms with the capability to thrive and propagate in exceedingly harsh conditions. These conditions are often detrimental to humans and the majority of known microorganisms, encompassing factors such as temperature, pH, and salinity parameters. Various extremophiles have been identified in environments marked by hydrostatic pressure, aridity, radiation, elevated temperatures, extreme pH levels, high salt concentrations, and substantial solvent/metal contents. Extensive documentation supports the notion that these microorganisms exhibit superior adaptation to extreme conditions compared to all other life forms on Earth. The capacity, or lack thereof, for organisms to acclimate to, endure, and reproduce within severe terrestrial settings provides valuable insights into life's tenacity in stressful scenarios. Extremophiles can be classified based on the physical attributes of the habitats they inhabit:

### *3.2.1 Thermophiles*

Over the past 40 years, thermophilic microorganisms have garnered significant attention among extremophiles (Bertoldo & Antranikian, 2002; Van der Maarel et al. 2002). These microorganisms possess the remarkable ability to flourish in elevated temperatures, ranging from 41 to 122 °C (Madigan & Marrs, 1997). They are categorized into hyperthermophiles (optimal growth temperature, $T_{opt}$ > or = 80 °C), extreme thermophiles ($T_{opt}$ between 70 and 80 °C), and moderate thermophiles ($T_{opt}$ between 45 and 70 °C), as commonly recognized definitions suggest (Mesbah & Wiegel, 2008; Wagner & Wiegel, 2008). Naturally heated environments on Earth span from terrestrial volcanic locations, such as solfatara fields, with temperatures slightly above ambient levels, to underwater hydrothermal systems (including sediments, submarine volcanoes, fumaroles, and vents) where temperatures surpass 300 °C. Other examples encompass subterranean sites like oil reservoirs and sun-warmed surface soils reaching temperatures of up to 65 °C. Additionally, there exist human-made heated environments like compost piles (typically around 60–70 °C but occasionally as high as 100 °C), industrial processes, slag heaps, and water heaters (Oshima & Moriya, 2008). Among geothermally heated habitats are alkaline hot springs with primarily carbonate content, typically at neutral pH, and acidic regions containing mud holes. In the majority of acidic, high-temperature habitats, elemental sulfur and metal sulfides are prevalent, with many isolates from these locales metabolizing sulfur through anaerobic respiration or fermentation.

The majority of hyperthermophilic organisms belong to the archaea domain, and many of these perform fundamental metabolic processes such as methanogenesis, anaerobic respiration involving sulfate reduction, sulfur reduction, nitrate reduction, iron reduction, etc., as well as aerobic respiration and even fermentation. Notable genera encompass Archaeoglobus, Thermodiscus, Thermoproteus, Acidianus, Pyrococcus, Thermococcus, Desulfurococcus, Methanothermus, Methanococcus, Methanopyrus, and nitrate-reducing Pyrobaculum and Pyrolobus. For a considerable time, *P. fumarii* held the record for the highest $T_{max}$ (113 °C) and a $T_{opt}$ of 106 °C, being unable to grow below 90 °C. However, the current record holder is the *M. kandleri* strain, isolated from the deep ocean near Japan, which boasts a remarkable $T_{max}$ of 122 °C under high atmospheric pressure (Takai et al., 2008).

Bacteria that thrive in extreme thermophilic conditions encompass a range of species, including anaerobic firmicutes, the cellulose-digesting *Caldicellulosiruptor saccharolyticus* (Rainey et al., 1994), the ethanol-producing *Thermoanaerobacterium ethanolicus* (Wiegel & Ljungdahl, 1996), and the facultative chemolithoautotrophic *Thermoanaerobacterium kivui* (Leigh & Wolfe, 1983). Recent discoveries have introduced novel thermophilic species with significance for both fundamental and practical scientific matters. Examples include the citrate-fermenting *Sporolituus thermophilus* (Ogg & Patel, 2009), the newly identified bacterial phylum Caldiseria (Mori et al., 2009), the deep-sea bacterium *Nautilia abyssi* (Alain et al., 2009), the thermal mud-inhabiting *Anoxibacillus thermarum* (Poli et al., 2009), and the novel microaerophilic, nitrate- and nitrite-reducing thermophilic bacterium *Microaerobacter geothermalis* (Khelifi et al., 2010).

Within the realm of moderate thermophiles and thermotolerant organisms, there are cellulose-digesting *Clostridium thermocellum*, acetogenic *Moorella thermoacetica/* thermoautotrophica, and Thermoanaerobacterium (formerly Clostridium) thermosaccharolyticum, known for growing in vacuum-sealed foods and nicknamed the "can-swelling" organism (Wiegel & Canganella, 2000; Prevost et al., 2010). Other notable species encompass the obligate mixotrophic *Thiomonas bhubaneswarensis*, the marine *Lutaonella thermophila*, cellulose-degrading bacteria *Clostridium clariflavum* and *Clostridium caenicola*, the facultative microaerophilic *Caldinitratiruptor microaerophilus*, and a newly discovered hydrogen-producing bacterium from buffalo dung (Arun et al., 2009; Panda et al., 2009; Shiratori et al., 2009; Fardeau et al., 2010; Romano et al., 2010).

Explorations into how these organisms maintain stability in extreme temperatures, particularly concerning protein and membrane structural levels, have shed light on various aspects of protein, membrane, and nucleic acid architecture. However, a comprehensive grasp of the principles underpinning thermostability remains incomplete (Daniel, 1996; Vieille & Zeikus, 2001; Basu & Sen, 2009; Kim et al., 2014; Averhoff & Müller, 2010). Advances in genetic tools tailored for these organisms hold promise for future practical applications (Bustard et al., 2000).

### 3.2.2 Psychrophiles

Psychrophiles and psychrotrophs are types of microorganisms specialized for cold environments. A psychrophilic prokaryote is defined by a temperature range for optimal growth below 15 °C and no growth above 20 °C. In contrast, psychrotolerant prokaryotes have optimal growth temperatures above 20 °C but can still thrive well below 5 °C (Bowman, 1997). Remarkably, over 80% of Earth's total biosphere maintains temperatures below 5 °C, encompassing environments like permafrost, cold rocks, Arctic and Antarctic ice, frigid sea water in polar regions, deep marine and freshwater sediments, as well as the bodies of cold-blooded organisms adapted to low temperatures.

Well-known genera containing various species of psychrotrophs include Pseudomonas, Flavobacterium, Achromobacter, Alcaligenes, Bacillus, Arthrobacter, Aeromonas, Serratia, Escherichia, Proteus, Psychroflexus, Psychromonas, Psychrobacter, and Vibrio. These microorganisms have been found in a multitude of cold habitats, even in the severe desert conditions of Antarctica (Friedmann, 1993). Additionally, newly discovered psychrophiles comprise Arctic bacteria such as *Marinobacter psychrophilus*, *Phaeobacter arcticus*, and *Moritella dasanensis*, as well as Antarctic bacteria including *Maribacter antarcticus*, *Sporosarcina antarctica*, and *Exiguobacterium soli*, and the methanol-utilizing methanogen *Methanolobus psychrophilus* (Kim et al., 2014; Zhang et al., 2008; Chaturvedi et al., 2008; Yu et al., 2008; Zhang et al., 2009).

Even cold-adapted yeast and fungi like Candida, Torulopsis, Penicillium, Aspergillus, and others have been identified. Remarkable isolates include the yeasts *Mrakiella cryoconiti* and *Rhodotorula hymalayensis* (Margesin & Fell, 2008; Shivaji et al., 2008). These organisms have been detected in permafrost and deep ice cores, suggesting their ability to endure exceedingly lengthy periods. However, laboratory studies have demonstrated that not all species can withstand the freezing and thawing process, especially during exponential growth phases. The capacity for growth at low temperatures appears to depend on membrane fluidity and the availability of liquid water. If an organism cannot adjust its membrane lipids, substrate transport across the cell membrane becomes compromised. The properties of the liquid inside and near the cell also play a role in freezing. Structural alterations in proteins have been observed, enabling enzymes to function in low-energy environments (Gerday et al., 1997). The relationship between the cold-shock response, encountered in both psychrophiles and mesophiles, and cold adaptation has also been explored (Russell, 1997; Ray et al., 1998). Despite exceptions, most psychrophiles inhabit nutrient-rich environments, possibly due to decreased uptake and transport system affinities at lower temperatures.

### *3.2.3 Acidophiles*

Acidophiles are a group of microorganisms that thrive and are uniquely adapted to acidic environments with low pH values, typically below pH 3. These extremophiles have evolved specialized strategies to withstand the challenges posed by high levels of acidity, which can include proton influx, metal toxicity, and disrupted cellular functions. Acidophiles encompass various organisms, including bacteria, archaea, and even some eukaryotic microorganisms, like certain fungi and algae. These organisms can be found in natural acidic environments such as acid mine drainage, volcanic springs, and acidic soils. Many acid-loving microorganisms, known as acidophiles, have developed highly effective strategies to maintain their internal cytoplasm at or around a neutral pH level. Various processes have been identified as contributing to pH balance in these acidophiles, as discussed in previous research by Dopson and colleagues (2004), and Baker-Austin and Dopson (2007). They often have proton-pumping ATPases and other mechanisms to maintain their internal pH and regulate proton influx. Acidophiles also employ strategies to counteract metal toxicity that can result from the dissolution of metals in acidic environments. These strategies may involve metal sequestration, efflux pumps, and complexation with organic molecules. Some acidophiles are chemolithotrophic, using inorganic compounds such as iron, sulfur, or hydrogen as energy sources. They play a crucial role in biogeochemical cycling by oxidizing minerals and releasing elements like iron and sulfur into the environment (Baker-Austin & Dopson, 2007).

In highly acidic environments where dissolved organic carbon is present in low quantities (<20 mg/l), acidophilic organisms can be classified as oligotrophic due to their metabolic preferences. One extensively studied prokaryote in this category is the bacterium *Acidithiobacillus ferrooxidans*, known for its iron and sulfur oxidation abilities (Leduc & Ferroni, 1994). Other organisms within this group include halotolerant Thiobacillus prosperous and *Leptospirillum ferrooxidans*. Among moderately thermophilic, acidophilic autotrophs, we find the bacterium *Leptospirillum thermoferrooxidans*. Among extremely thermophilic archaea, various species of Acidianus, *Metallosphaera sedula*, and *Sulfurococcus yellowstonii* thrive. Presence of light in extremely acidic environments, favours phototrophic microorganisms as primary producers. This includes eukaryotic algae like Euglena spp., Chlorella spp., *Chlamydomonas acidophila*, *Ulothrix zonata*, and *Klebsormidium fluitans* (Gyure et al., 1987; Lopez-Archilla et al., 1995), as well as the single-celled rhodophyte *Galdieria sulphuraria* (Brock, 1978). Among the mixotrophic or heterotrophic acidophiles are iron oxidizers like *Ferrimicrobium acidiphilum* (a mesophilic obligate heterotroph), *Sulfobacillus acidophilus, Sulfobacillus thermosulfidooxidans*, and *Acidimicrobium ferrooxidans*. Sulfur oxidizers within this group encompass organisms such as mesophiles *Acidiphilium acidophilus*, *Thiomonas cuprina*, and *Sulfobacillus disulfidooxidans*; extreme thermophiles like *Sulfolobus shibitae, Sulfolobus solfataricus*, and *Sulfolobus hakonensis*; and *Sulfurococcus mirabilis* and *Metallosphaera prunae*.

The unique characteristics of acidophiles have practical applications. For instance, they can be harnessed for bioleaching, a process that uses microbial metabolism to extract metals from ores. Acidophiles are also employed in bioremediation to remove heavy metals from contaminated sites. Additionally, their enzymes and biomolecules, adapted to function in acidic conditions, have potential industrial and biotechnological applications (Johnson & Hallberg, 2008).

### 3.2.4 Alkaliphiles

The term "alkaliphile"is used to describe microorganisms that exhibit optimal or robust growth at pH levels exceeding 9, while their growth is either hindered or sluggish at the nearly neutral pH of 6.5. Alkaliphiles can be sourced from ordinary environments like garden soil, though the prevalence of viable alkaliphiles is more prominent in samples originating from alkaline surroundings. The outer cell surface appears to play a pivotal role in maintaining the intracellular pH within the range of 7–8.5, enabling alkaliphiles to flourish in alkaline habitats. Nonetheless, the mechanisms underlying these adaptation processes remain unclear. Naturally occurring highly alkaline environments are found in "soda lakes," which result from high rates of evaporation within enclosed drainage basins. These lakes contain minimal amounts of Mg2+ and Ca2+ ions and are nearly saturated with sodium salts, particularly chloride, carbonate, and bicarbonate. As a result, the pH typically hovers around 10 due to elevated levels of sodium carbonate (Duckworth et al., 1996). Alkaline conditions are also present in localized areas of regular soils, where temporary alkalinity is attributed to various biological activities (Grant et al., 1990).

Several natural habitats are both hypersaline and alkaline, providing a suitable environment for a diverse group of haloalkaliphiles. Notably, Lake Magadi in Kenya, the Wadi Natrun Lakes in Egypt (Mesbah et al., 2009), and the soda lakes of Kulunda Steppe in Altai, Russia (Sorokin & Muyzer, 2010a, b) host these organisms. Artificial alkaline environments include discharges from tanneries, paper mills, food and textile processing facilities, calcium carbonate kilns, detergents, and other industrial processes. Within the realm of alkaliphilic prokaryotes, there are distinctive groupings. Among the cyanobacteria, certain members of the

Chroococcales order and several species within the Spirulina genus are alkaliphiles. The aerobic Firmicutes group encompasses widespread alkaliphiles, including those from the Actinomycetales order, the Micrococcaceae families, the Streptomyces and Nocardia genera, as well as related actinomycetes. Alkaliphiles are also found in various genera within the Bacillales order. The Proteobacteria phylum houses many alkaliphiles, particularly within the gamma subdivision, including Ectothiorhodospira, Halomonadaceae, and Pseudomonas. Among the Archaea, numerous alkaliphiles are found in the Euryarchaeota group, such as *Halorubrum vacuolatum, Natrialba magadii*, several Natronobacterium species, and several Natronococcus species. Alkaliphilic Methanomicrobiales comprise several Methanohalophilus species.

Alkaliphiles adeptly maintain their internal pH levels close to neutrality despite the alkaline external surroundings. This phenomenon has been substantiated by subsequent observations (Padan et al., 2005), demonstrating that aerobic alkaliphiles generally maintain their cytoplasmic pH approximately 2 units lower than the ambient pH. They employ diverse strategies, including expelling excess protons, utilizing alkaline shock proteins, and possessing altered cell membranes with reduced proton permeability. Alkaliphile cell membranes might feature elevated quantities of sphingolipids and exhibit distinct lipid compositions in contrast to organisms that thrive in neutral conditions. This attribute aids in sustaining membrane integrity and functionality within alkaline settings. Some alkaliphiles accumulate specific ions, such as sodium ions, to counterbalance the electrochemical gradient across their membranes, facilitating pH homeostasis (Ito & Guffanti, 2002). Many alkaliphile membranes are reported to incorporate sodium-coupled ATPases (similar to those found in halophiles) in addition to, or sometimes instead of, proton-coupled ATPases (Koyama, 1999; Ueno et al., 2000); however, conflicting data exists (Cavicchioli & Thomas, 2000).

### *3.2.5 Barophiles*

Barophiles, also known as piezophiles or pressure-loving microorganisms, are a unique group of extremophiles that thrive in high-pressure environments, specifically in the deep ocean and beneath the ocean floor (Lauro & Bartlett, 2008). These organisms have evolved to withstand the tremendous pressures found at great depths, where the hydrostatic pressure can exceed 1000 times that of atmospheric pressure. Barophiles offer insights into the adaptations, life forms can make to survive in extreme conditions and have implications for astrobiology, biotechnology, and environmental science. These adaptations often involve modifications to cellular membranes, enzymes, and other cellular components that can be affected by pressure changes. For instance, their membranes may have unique lipid compositions that provide greater stability under high pressures. Enzymes found in barophiles often display increased flexibility, allowing them to function effectively under extreme pressures. Additionally, these organisms may possess specialized proteins that protect cellular structures from pressure-induced damage (Bartlett, 2002).

Barophiles are predominantly found in the deep ocean, where the pressure increases with depth. They inhabit regions such as deep-sea trenches, hydrothermal vents, and sub-seafloor sediments. One remarkable habitat for barophiles is the hadal zone, which includes ocean trenches with depths greater than 6000 meters (Tamegai et al., 2017). In these environments, they coexist with other extremophiles like thermophiles and psychrophiles, adapting to specific niches within the challenging conditions of the deep sea. Understanding barophiles is not only essential for deciphering the limits of life on Earth but also has broader implications. Their adaptations could potentially offer insights into life's possibilities in extreme conditions on other planets or celestial bodies. Additionally, their unique enzymes and biomolecules have practical applications in biotechnology, as they might function effectively under high-pressure conditions used in industrial processes (Scoma & Boon, 2016). Barophiles also play a role in biogeochemical cycles and influence nutrient cycling in the deep ocean.

### 3.2.6 Radioresistant Organisms

Radiation-resistant organisms, also known as radioresistant organisms or radiotolerant organisms, are living entities that possess the remarkable ability to survive and thrive in environments with high levels of ionizing radiation. These organisms have evolved strategies to withstand the damaging effects of radiation, which can break DNA strands, induce mutations, and cause cellular damage. Studying radiation-resistant organisms provides insights into the mechanisms of DNA repair, cellular protection, and adaptation to extreme conditions. Radiation-resistant organisms can withstand various forms of ionizing radiation, including gamma rays, X-rays, and cosmic radiation. These types of radiation have enough energy to ionize atoms and molecules, leading to the creation of reactive free radicals that can damage cellular components, particularly DNA (Daly, 2009).

Radiation-resistant organisms are found in diverse environments, such as high-altitude regions, space, deep-sea hydrothermal vents, nuclear power plants, and radioactive waste disposal sites (Cox & Battista, 2005). Notable examples include extremophiles like *Deinococcus radiodurans*, which was initially isolated from irradiated meat, and certain bacterial species in the genus Thermococcus, which inhabit hydrothermal vents on the seafloor. These organisms have evolved several mechanisms to survive high radiation levels. One of the most studied mechanisms is the ability to repair DNA damage (Daly, 2012). For instance, *Deinococcus radiodurans* possesses multiple copies of its genome, allowing it to recover from extensive DNA damage by reassembling intact copies of its genome. Other strategies include efficient DNA repair enzymes, protective proteins that shield cellular components from radiation, and antioxidants that counteract oxidative stress induced by radiation (Battista & Cox, 2005). Besides, *Deinococcus radiodurans*, other radiation-resistant organisms include certain species of Thermococcus, *Rubrobacter radiotolerans*, and some fungi like *Cryptococcus neoformans* (Dadachova & Casadevall 2007; Battista & Cox, 2005).

DNA repair processes in radiation-resistant organisms are particularly efficient. They can rapidly repair double-strand breaks and other types of DNA damage caused by ionizing radiation. Enzymes involved in these repair processes include DNA polymerases, exonucleases, and recombinases. The unique abilities of radiation-resistant organisms have potential applications in biotechnology and medicine. Their DNA repair mechanisms, for example, could be employed to enhance the radiation resistance of other organisms or to protect DNA-based materials from damage during irradiation (Makarova et al., 2007). Understanding radiation-resistant organisms has implications for astrobiology and space exploration. These organisms offer insights into how life might persist in radiation-rich environments on other planets, and they could inspire strategies to protect astronauts from space radiation during long-duration missions (Daly, 2009).

## 3.3 Extremozymes: Types and Functions

Extremophilic microorganisms are a valuable source of extremozymes, which exhibit a diverse array of applications in various industries due to their impressive stability and biodegradability (Jaenicke et al., 1996; Sthal, 1993). These extremozymes, functioning effectively under extreme environmental conditions previously considered incompatible with biological activity, have expanded the horizons of industrial biocatalysis. Through their application, a wide spectrum of resilient biomolecules has become available for industrial use, encompassing cold-tolerant, acid-tolerant, alkali-tolerant, and salt-tolerant extremozymes (Cavicchioli et al., 2002). This paradigm shift has enabled the deployment of biologically active agents in environments once deemed unsuitable. The ongoing pursuit of enzyme research emphasizes the exploration of novel extremozymes characterized by unique and enhanced capabilities, further propelling the boundaries of industrial applications (Raddadi et al., 2015). Here, we delve into the types and functions of enzymes found in extremophiles:

### *3.3.1 Thermophilic Enzymes*

Thermophilic enzymes constitute a subgroup within extremophilic enzymes that exhibit both activity and stability at high temperatures. They exhibit unique adaptations that enable them to maintain their structure and function even under extreme temperature conditions. Additionally, they possess the capability to enhance the accumulation of hydrophobic residues and form disulfide bonds between oppositely charged ions (Brasen et al., 2005; Mayer et al., 2012). These enzymes have garnered substantial interest due to their potential applications in various industrial processes, particularly those necessitating elevated temperatures. A diverse array of enzymes derived from thermophilic microorganisms has been extensively characterized. This array encompasses cellulases, amylases, pullulanases, xylanases, mannanases, pectinases, chitinases, proteases, lipases, esterases, and phytases (Sunna & Bergquist, 2003). This versatile range of enzymes holds promise for their role in driving various industrial applications.

Thermophilic proteases are active at elevated temperatures and play a role in protein hydrolysis. They have applications in the food industry for meat tenderization and protein modification (Gupta et al., 2002). Amylases catalyze the breakdown of starch into simpler sugars at high temperatures, making them valuable in the starch and ethanol industries. They are crucial for industries like starch processing, where they enhance the efficiency of saccharification processes at high temperatures (Kuddus & Singh, 2012). Polymerase enzymes are a cornerstone of molecular biology, as they are used in polymerase chain reactions (PCR) for DNA amplification. They can withstand the denaturation temperatures required for DNA strand separation (Kaledhonkar et al., 2016). Thermophilic lipases are active and stable at high temperatures and can catalyze various lipid-related reactions, including hydrolysis and esterification (Jaeger & Eggert, 2002). They are being used in different processes like esterification, interesterification, transesterification, grease hydrolysis and organic biosynthesis. Thermophilic Cellulases are enzymes that hydrolyze cellulose into glucose units and have applications in biofuel production, as they can efficiently degrade cellulose at elevated temperatures (Gaur et al., 2015).

### *3.3.2 Psychrophilic Enzymes*

Most psycrophilic enzymes are cold active and thermolabile, and these characteristics focuses its biotechnological interest. They have evolved to function optimally under such frigid conditions, displaying adaptations that enable them to maintain their functionality in environments where higher temperatures might denature typical enzymes. An important benefit of cold-adapted enzymes lies in their ability to function effectively without the necessity of maintaining large-scale processes at higher temperatures. This unique attribute has the potential to reduce both the carbon footprint and energy expenditures associated with biocatalysis on a significant scale. However, one drawback of psychrophilic proteins is their susceptibility to temperature-induced instability, limiting their suitability for industrial catalysis applications. Genetic engineering approaches are often employed to address these limitations, enhancing their stability or activity at low temperatures and even modifying their pH preferences and other biochemical characteristics (Feller, 2003).

Currently, among cold-adapted enzymes, only a couple of lipases have found application in industrial settings. These include lipases from *Yarrowia lipolytica* and *Candida antarctica* lipase B (CalB), with CalB immobilized on an acrylic resin known as Novozyme 435. In molecular biology, one prominent psychrophilic enzyme is alkaline phosphatase, which plays a crucial role in DNA modification processes, such as dephosphorylating DNA vectors before cloning and removing phosphate groups from the 5′ end of DNA strands for further manipulation. Florczak et al. (2013) discovered a cold-adapted lipase, LipG7, derived from the Antarctic filamentous fungus Geomyces sp. P7. This enzyme retained 15% of its maximal activity even at 0 °C. LipG7 demonstrated potential for industrial use as an enantioselective biocatalyst, effectively catalyzing enantioselective transesterification of secondary alcohols.

Psychrophilic enzymes have garnered significant attention due to their potential applications in industries requiring low-temperature processes. These enzymes have the capacity to operate efficiently at temperatures near or even below freezing points, making them invaluable tools in fields such as food preservation, bioremediation of cold environments, and the production of cold-active detergents. Numerous enzymes from psychrophilic microorganisms have been investigated and characterized, including amylases, proteases, lipases, cellulases, and more. These enzymes not only hold promise for industrial processes but also offer insights into the molecular adaptations that enable them to function effectively in cold habitats.

### *3.3.3 Acidophilic and Alkaliphilic Enzymes*

Acidophiles are organisms that thrive in environments with an optimal pH level below 3–4 (Jaenicke, 1981). However, the mechanisms underlying the adaptation of acidophilic proteins remain relatively unexplored. For instance, the endo-β-glucanase enzyme from *Sulfolobus solfataricus* has exhibited stability at its optimum pH of 1.8 (Huang et al., 2005). Nevertheless, this enzyme struggles to maintain its stability in highly acidic environments, particularly at pH values below 2 (Golyshina & Timmis, 2005). The α-glucosidase enzyme derived from *Ferroplasma acidiphilum* has showcased stability under low pH conditions. Interestingly, this enzyme has displayed a preference for a pH of 3 over the expected internal pH of *Ferroplasma acidiphilum*, which is 5.6 (Sharma et al., 2012). Additionally, similar pH optima have been observed for other cytoplasmic enzymes, all of which exhibit significantly lower activity levels at pH values exceeding 5. Another example of an acidophilic enzyme is the acid protease derived from the acidophilic fungus *Aspergillus niger*. This enzyme exhibits optimal activity in acidic conditions and finds applications in food and beverage industries for the production of various products, including dairy and meat alternatives (Kumar & Goyal, 2019).

Cellulolytic and xylanolytic enzymes have garnered attention due to their crucial role in breaking down complex polysaccharides. In specific industrial processes, these enzymes operate within an acidic environment characterized by high temperature and acidity. Their presence facilitates the hydrolysis of cellulose-rich materials, a pivotal step in converting biomass into biofuels and ethanol. This enzymatic hydrolysis process breaks down the intricate cellulose and hemicellulose structures into simpler sugars that can be subsequently fermented to produce biofuels (Jackson et al., 2007). It is noteworthy that acidophilic enzymes tend to form multienzyme complexes near the pH optima resembling the cytoplasmic environment (Hauenstein et al., 2004; Szilágyi & Závodszky, 2000). By harnessing the capabilities of acidophilic enzymes, researchers and industries can optimize the efficiency of biofuel and ethanol production processes. The unique adaptations of these enzymes to extreme pH and temperature conditions make them valuable tools for sustainable and environmentally friendly approaches to energy production.

Alkaliphilic enzymes are a remarkable class of enzymes originating from microorganisms that flourish in alkaline environments with high pH levels. These enzymes have evolved specific adaptations to function optimally under these extreme alkaline conditions. Their structural and biochemical characteristics enable them to maintain stability and catalytic activity even at elevated pH values. Alkaliphilic enzymes have attracted considerable attention due to their potential applications across diverse industrial sectors and biotechnological fields.

One prominent example of an alkaliphilic enzyme is alkaline protease, often derived from bacteria of the Bacillus genus. These proteases possess distinct features that allow them to perform effectively under alkaline conditions, making them suitable for applications in various industries. The enzyme's stability and activity at higher pH levels make it particularly valuable in the production of laundry detergents, where it aids in removing protein-based stains. Additionally, alkaline proteases are used in leather processing to soften leather and enhance its quality (Bhaskar et al., 2007). Alkaline cellulases are another noteworthy subgroup of alkaliphilic enzymes. These enzymes are capable of efficiently breaking down cellulose, a complex polysaccharide found in plant cell walls. Alkaline cellulases find application in various sectors, including biofuel production, where they assist in the conversion of lignocellulosic biomass to bioethanol. Moreover, they are used in textile processing to remove impurities and aid in fabric dyeing, as well as in the paper industry to facilitate pulp degradation (Sukumaran et al., 2005). Research and exploration into alkaliphilic enzymes continue to yield new insights and expand their potential applications. As industries seek sustainable and efficient solutions, the unique properties of these enzymes offer promising avenues for innovation in fields such as waste management, bioremediation, and the production of valuable bio-based products.

### 3.3.4 Halophilic Enzymes

Enzymes derived from halophilic microorganisms have developed distinct adaptation strategies that contribute to their remarkable stability under conditions of low water activity and the presence of organic solvents (Datta et al., 2010). Extremozymes originating from halophiles encompass a diverse array of enzymes such as xylanases, amylases, proteases, and lipases. These enzymes are produced by a range of halophilic microorganisms including Acinetobacter, Haloferax, Halobacterium, Halorhabdus, Marinococcus, Micrococcus, Natronococcus, Bacillus, Halobacillus, and Halothermothrix, as documented by Sutrisno et al. (2004) and Taylor et al. (2004). Lipases and esterases, in particular, hold significant potential for industrial applications, notably in the realms of polyunsaturated fatty acid production, food processing, and biodiesel manufacture (Litchfield, 2011).

Halophilic α-amylases hydrolyze starch into smaller sugars, and they are of great interest due to their stability and activity in high-salt conditions. These enzymes find applications in starch processing, where they aid in the production of maltodextrins and other starch-derived products. One case study demonstrated the isolation and characterization of a halophilic α-amylase from the bacterium *Halomonas meridiana*. This enzyme displayed remarkable stability and activity in the presence of salt, making it valuable for industrial applications (Coronado et al., 2000). Pérez-Pomares (2003) studied production of a halophilic protease from the archaeon *Haloferax mediterranei*. The enzyme showed excellent stability and activity under high-salt conditions, making it suitable for protein hydrolysis in industrial processes. Halophilic lipases have potential applications in various industries, including food processing, biodiesel production, and the detergent industry. Isolation and characterization of a halophilic lipase from the bacterium *Bacillus licheniformis* has exhibited stability and activity in the presence of high salt concentrations, making it a promising candidate for lipid hydrolysis in industrial settings (Kumar et al., 2012). Halophilic DNA polymerases play a crucial role in DNA replication and amplification. These enzymes have applications in molecular biology techniques such as polymerase chain reaction (PCR). Baliga et al., (2004) investigated the use of a halophilic DNA polymerase from the archaeon *Haloarcula marismortui* for DNA amplification in high-salt PCR conditions.

Up to this point, research focusing on the enzymatic behavior of halophilic enzymes within non-aqueous environments remains relatively limited. For instance, investigations have been carried out on proteases sourced from organisms like *Halobacterium halobium*, *Saliniovibrio* sp. strain AF-2004, *Natrialba magadii* and *Salimicrobium halophilum* strain LY20, as well as glutamate dehydrogenase from *Halobacterium salinarum* strain NRC-36014 (Karbalaei-Heidari et al., 2007; Ruiz & De Castro, 2007; Fukushima et al., 2005; Shafiei et al., 2011). However, the biotechnological applications of halophilic enzymes extend beyond their stability under high salt concentrations. These enzymes also exhibit tolerance to elevated temperatures and remain stable even in the presence of organic solvents (Zaccai, 2004). Their functionality is sustained in environments characterized by low water activity, as they maintain sufficient water content to facilitate a suitable charge distribution at the active site, thereby preserving the enzyme's conformation (Sellek & Chaudhari, 1999). The effectiveness of halophilic enzymes is attributed to various mechanisms that contribute to their stability and solubility when faced with high concentrations of sodium chloride and potassium chloride, including interactions with organic solvents and structural features embedded within the three-dimensional enzyme structure (Cordone et al., 1999).

### 3.3.5 Barophilic (Piezophilic) Enzymes

Piezophiles are organisms that have aptly adapted to flourish in environments characterized by high hydrostatic pressures, specifically within deep-sea and volcanic regions. Notable examples include *Pyrococcus abyssi* (Boonyaratanakornkit et al., 2002; Fang et al., 2010; Reed et al., 2013). These organisms exhibit optimal growth conditions under pressures surpassing atmospheric levels (Simonato et al., 2006). Their prevalence spans various phylogenetic groups, implying multiple instances of independent piezophilic evolution. However, piezophiles remain among the less comprehended extremophiles, largely due to the challenges associated with sample collection and experimentation under extreme pressures. The adaptation of piezophiles has been elucidated through studies involving proteins like Sso7d from *Sulfolobus solfataricus*, highlighting their unique characteristics (Fusi et al., 1995; Mombelli et al., 2002). For instance, piezophilic proteins such as the peptidase sourced from *Pyrococcus horikoshii* exhibit considerable stability under elevated pressure. The potential of piezophilic enzymes has been expounded in reviews by Cavicchioli et al. (2002) and Georlette et al. (2004). It's worth noting that unlike thermopiezophiles, piezophilic microorganisms lack specialized channels for stability.

An exemplar of a barophilic enzyme is the α-amylase derived from the deep-sea bacterium *Photobacterium profundum*. This enzyme has demonstrated heightened efficacy and stability under high pressures, rendering it a promising contender for industrial applications necessitating elevated pressures (Bartlett, 2002). Despite the considerable potential of piezophilic enzymes in industrial contexts, research in this domain remains relatively sparse. Abe and Horikoshi (2001) showcased the distinctive capabilities of α-amylase from piezophilic proteins, showcasing the production of trisaccharides from maltooligosaccharide substrates at high pressures with minimal energy expenditure. This reaction holds substantial promise for industrial and biotechnological applications, particularly within the realm of the food industry (Cannio et al., 2004). Piezophilic proteins have effectively demonstrated remarkable efficiency in an array of industries, encompassing detergents, food processing, and chemical products, underscoring their capacity to contribute significantly to the advancement of various industrial processes.

## 3.4 Industrial Applications of Extremophile Enzymes

The utilization of extremophilic enzymes extends beyond industrial processes to include environmental and medical applications. However, their exploitation requires careful consideration of ethical and environmental implications to ensure sustainable practices. By harnessing the diverse enzymatic capabilities of extremophiles, researchers and industries can unlock innovative solutions for biotechnology, bioengineering, and more. Figure 3.1 illustrates various extremozymes and their industrial applications.

### *3.4.1 Biofuel Production*

Over the last decade, there has been a growing emphasis on biofuels as a viable and environmentally friendly substitute for traditional fuels (Luque et al., 2008). Within a few decades, biofuel production and consumption have doubled in several countries (Hoekman, 2009). Biofuels are products of fermentation and microbial metabolism and traditionally include ethanol, butanol, methane,

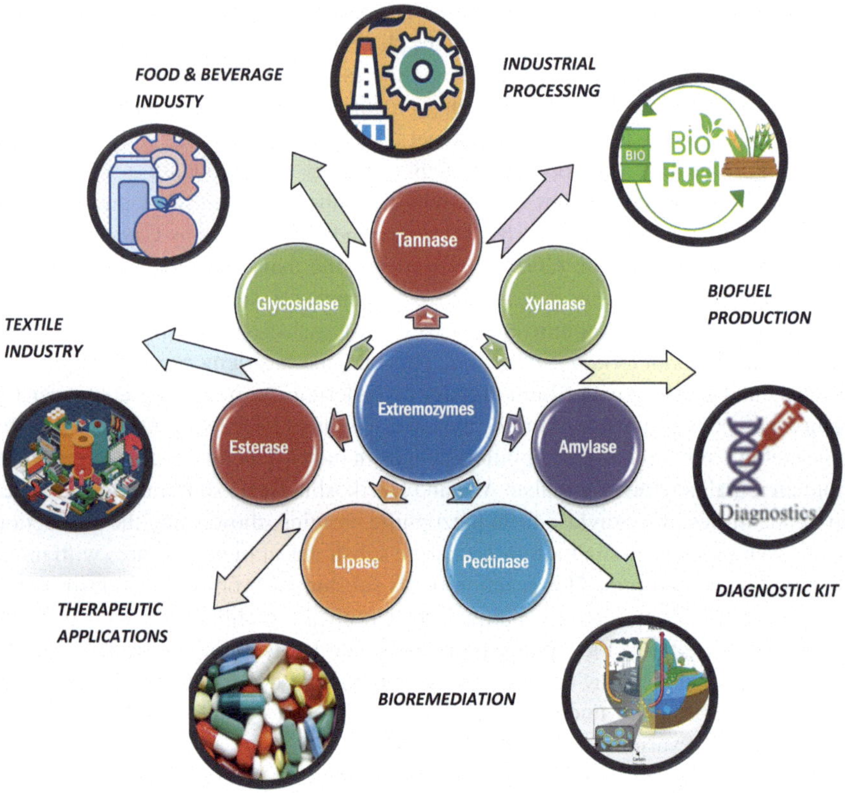


**Fig. 3.1** Extremozymes and their industrial applications

hydrogen, and long-chain alkyl ester derivatives. Largely, biofuels have been produced using raw materials like sugar cane, corn, beets, wheat, sorghum, rapeseed for years using conventional processing techniques. Recently, the advent of biofuel processing has begun with lignocellulosic materials (Naik et al., 2010). Enzymes derived from extremophiles, particularly cellulases, hemicellulases, and lignin-degrading enzymes, play a pivotal role in breaking down the intricate structure of lignocellulosic biomass into fermentable sugars. These sugars are then utilized in the production of biofuels such as bioethanol and biohydrogen. The advantage of extremophilic enzymes lies in their enhanced stability and activity at elevated temperatures, enabling efficient biomass conversion (Siddiqui & Cavicchioli, 2006).

Among these enzymes, lipolytic enzymes and amylases (both $\alpha$ and $\beta$) are prominent extremozymes. Noteworthy examples of thermostable lipases include those from *Streptomyces bambergiensis* OC 254, Microbacterium sp., Bacillus sp. THL027, and *Geobacillus thermodenitrificans* all exhibiting potential for biodiesel synthesis (Ugur et al., 2014; Tripathi et al., 2014; Dharmsthiti & Luchai, 1999; Christopher et al., 2015). Industrially, amylases find utility in cleaving 1,4-$\alpha$-D-glucosidic linkages within amylose chains sourced from starch.

Numerous sources of thermostable amylases have been identified, including *Bacillus licheniformis, Bacillus* sp. AB68, *Bacillus* sp. A315, *Bacillus cohnii, Bacillus* sp. AK-2, Geobacillus sp., and Anoxybacillus sp. AH1, each with distinct enzymatic properties (Fitter et al., 2001; Aygan et al., 2008; Ghorbel et al., 2009; Srivastava et al., 1984; Febriani et al., 2019; Zhu et al., 2020).

Del Campo et al., (2013) harnessed the extremozyme polyphosphate xylulokinase from the hyper-thermophilic *T. maritima* to enhance bioethanol production. Similarly, Zhang et al. (2015) employed *T. maritima* for extracting another extremozyme, β-1,4-endoglucase, for ethanol production from carboxymethyl cellulase-Na. de Camargo et al. (2018) contributed to second-generation bioethanol production using arabinofuranosidase from *C. thermocellum*, exhibiting an enzyme activity of 184 U/mg. Kim et al. (2014) explored continuous immobilized crosslinked yeast fermentation for bioethanol production, achieving an 89% ethanol conversion rate from micro-algal *Chlorella vulgaris* hydrolysate. Comparatively, Borges et al. (2014) evaluated immobilized and free state β-glucosidase enzymes for bioethanol conversion, highlighting the enhanced thermal stability offered by immobilization.

### 3.4.2 Bioremediation

Chemical compounds such as polychlorinated biphenyls and polyaromatic hydrocarbons (PAHs) exhibit toxic, mutagenic, and carcinogenic properties, posing significant health hazards to humans due to their release and accumulation in the environment (Yadav et al., 2016). Addressing this issue, researchers are increasingly concerned about the need for effective, environmentally friendly, and cost-efficient methods to remediate these pollutants.

In this context, extremophilic enzymes have emerged as crucial agents in bioremediation efforts, breaking down complex pollutants in harsh environments such as contaminated soils and waters. Enzymes sourced from extremophiles are better equipped to degrade pollutants, including hydrocarbons and xenobiotics, particularly under extreme conditions (Aono & Horikoshi, 1983). Prominent microbial species like Chloroflexi, Deinococcus-Thermus, and Chlorobi have demonstrated proficiency in reducing arsenite concentrations in the environment (Engel et al., 2013). Other bacteria, such as Geobacillus, hold potential for degrading organic pollutants (Orellana et al., 2018). Notably, halophilic bacteria generate substantial quantities of biosurfactants, amplifying the biodegradation of petroleum hydrocarbons (Kumar et al., 2007).

Relevant studies have highlighted specific enzymes for pollutant degradation. For instance, *Bacillus safensis* (CFA-06) produces oxidoreductase, which finds applications in degrading aromatic compounds (Da Fonseca et al., 2015). Other investigations have reported the extraction of laccase enzymes from *Thermus thermophilus* HB27, aiding in polyaromatic hydrocarbon degradation, as well as phosphotriesterase from *Geobacillus stearothermophilus*, contributing to organophosphate degradation (Peeples, 2014). Similarly, haloalkane dehalogenases from *Pseudomonas stutzeri* DEH130 and *Alcanivorax dieselolei* strain B-5 have shown promise in degrading halogenated pollutants (Di Donato et al., 2019).

White-rot fungi, such as MnP Irpex lacteus (Baborova et al., 2006) and *Pleurotus ostreatus* Pozdnyakova et al. (2006), have been identified as agents capable of degrading PAHs. Additionally, laccases, known for oxidizing xenobiotics and non-phenolic compounds, have applications in the degradation of dyes like remazol black-5, remazol blue-19, and remazol orange-16 (Viswanath et al., 2008).

In the context of pollution incidents, Nocardioides sp. strain KP7, isolated from a Kuwait beach, has demonstrated the capability to produce dioxygenase enzymes for the degradation of phenanthrene post an oil spill (Saito et al., 2000). For wastewater containing oil, lipases have proven valuable in biodegradation, even in extreme environments (Sharma et al., 2011), with reports of lipase enzymes from Moraxella sp. and yeast strains aiding in oil degradation (Wang et al., 2007). Furthermore, these enzymes are instrumental in the treatment of saline wastewater and oilfield waste. Ester-containing pesticides, including carbamates, organophosphates, and pyrethroids, are effectively targeted by microbial esterases, which break down these compounds into non-toxic forms. A diverse range of microbes has demonstrated the ability to transform pesticides into environmentally acceptable forms (Bhatt et al., 2016), with associated gene expression changes induced by pesticides observed in agricultural microbial communities (Bhatt et al., 2019). Microbes from cold regions are harnessed for bioremediation, and their enzymes and whole cells have applications in psychrophilic biofertilizer production. Industries dealing with phenolic compounds, like those involved in coal conversion, plastic, and resin manufacturing, and petroleum refining, face challenges in treating their wastewater (Lee et al., 1996). Enzymatic methods have proven effective since the 1980s, with horseradish peroxidase being employed to oxidize phenol Klibanov et al. (1983). The β-tyrosinase enzyme from the thermophile Symbiobacterium sp. SMH-1 has demonstrated the potential for converting phenol, often found in phenolic resin manufacturing, into the valuable product L-tyrosine (Lee et al., 1996).

### 3.4.3 Food and Beverage Industry

Extremophilic enzymes find applications in the food and beverage industry due to their stability and activity across a wide temperature range. Thermophilic and hyperthermophilic enzymes are employed in processes such as starch hydrolysis, protein modification, and fruit juice clarification, contributing to the production of various food and beverage products (Feller & Gerday, 2003). Thermophilic amylases find significant utility in the breakdown of gelatinized starch and the process of saccharification. While currently employed bacterial thermostable α-amylases have their industrial applications (Declerck et al., 2003), the demand persists for amylases that exhibit stability across diverse pH and temperature conditions. Extremophilic amylases, such as those derived from Thermus aquaticus, have demonstrated their efficacy in starch hydrolysis for various food products, including syrups, sweeteners, and maltodextrins. These extremophilic enzymes remain active even at high temperatures during starch gelatinization and cooking, thereby enhancing production yields and streamlining processing times (Bajpai, 2016).

Research has unveiled a range of extremozymes categorized as thermophilic, psychrophilic, and salt-tolerant amylases, originating from various bacterial genera, such as Desulfurococcus, Pseudoalteromonas, Pyrococcus, Rhodothermus, and Thermococcus (Duffner et al., 2000; Gomes & Steiner, 2004). Among these, thermophilic alkaline proteases contribute to the breakdown of excess meat adhering to bones (Chaplin, 2015). Conversely, cold-active proteases not only enhance meat flavors and tenderness during refrigeration but also play a role in hydrolyzing protein turbidity complexes in cold-stored fruit juices. Heat-resistant proteases, on the other hand, aid in improving the nutritional quality and functionality of protein hydrolysates (Synowiecki, 2010). However, it's worth noting that despite the characterization of several thermostable bacterial and fungal proteases, challenges in their downstream processing and the maintenance of their unique properties during large-scale production have led to increased costs for their industrial utilization. Meanwhile, the potential of low-temperature active enzymes remains significant (Feller & Gerday, 2003; Georlette et al., 2004). For instance, cold-active b-galactosidase serves the dairy industry in the preparation of lactose-free milk and cheese. Proteases sourced from psychrophilic extremophiles are harnessed in the dairy sector, particularly in cheese production. These enzymes effectively degrade milk proteins at lower temperatures, thus obviating the need for energy-intensive heating steps. This not only enhances cheese flavor and texture but also preserves its nutritional content (Gänzle, 2015). Similarly, in the bakery industry, glycosidases with low-temperature activity, including amylases, proteases, and xylanases, contribute to improved texture and flavor. Additionally, low-temperature active pectinases assist in fruit processing by retaining flavor and aiding in clarification. Tannases, enzymes catalyzing the hydrolysis of gallic acid esters and hydrolysable tannins, find applications in removing tannins from beverages like green tea and in instant tea production (Yao et al., 2014). Xylanases, primarily used in baking and feed additive industries, have garnered interest in the functional food sector due to their ability to synthesize xylo oligosaccharides and ferulic acid. Diverse extremophilic xylanases with distinctive properties have been reported in various metagenomic libraries (Cheng et al., 2012; Ziemer, 2013). Additionally, cellulases are utilized in animal feed processing to enhance forage nutritive value and cereal digestibility (Dhiman et al., 2002).

Alkaline lipases sourced from alkaliphilic extremophiles are chiefly employed in the detergent industry for their adeptness in effectively removing fatty stains. Moreover, they find application in the baking sector to modify fats used in bakery products, thereby enhancing texture and extending shelf life (Hasan et al., 2006). In a different context, halophilic extremozymes contribute to the production of high-salt content foods, imparting unique flavors and textures to items like pickles and fermented products. These enzymes retain their activity even in saline environments, making them well-suited for such applications (Ventosa et al., 1998). Furthermore, halophilic proteases have been identified as advantageous in fermentation processes (Vidyasagar et al., 2009). In the beverage industry, extremophiles that can operate under extreme pH conditions are employed to clarify fruit juices by breaking down pectins and proteins, thereby enhancing the clarity and stability of the final product (Gupta & Gigras, 2014).

Within the realm of pectinases, which include pectolyases, pectozymes, and polygalactouranase, classification under the GH28 glycoside hydrolases family is noteworthy. Widely employed in the fruit juice industry, these enzymes, in combination with plant cell wall degrading enzymes, contribute to the production of essential oils from substances like pepper and cardamom (Chandran et al., 2012). Moreover, Singh et al. (2012) reported a novel thermostable polygalacturonase derived from soil metagenomes, characterized for its unique biochemical properties. Similarly, Sathya et al. (2014) isolated functional genes of polygalacturonase with comparable properties from a soil metagenomic library. Incorporating extremozymes into the food and beverage industry offers several benefits, including increased process efficiency, reduced energy consumption, improved product quality, and the development of novel food products. These applications highlight the versatility and potential of extremozymes to revolutionize various aspects of food and beverage production.

### 3.4.4 Pharmaceutical Industry

The growing demand within the biotech industries and the emergence of novel industrial processes have acted as significant catalysts for the initiation of extremophilic research. Consequently, enzymes derived from extremophilic sources have exerted a substantial influence, particularly from both commercial and biotechnological standpoints. Extremozymes can catalyze specific reactions under extreme conditions, such as high temperatures, extreme pH levels, or high salinity. This ability can be harnessed in drug development, enabling the synthesis of complex molecules that might be challenging using traditional methods. For example, extremozymes can be used to create chiral molecules with high enantioselectivity, a critical factor in pharmaceutical synthesis. They can help maintain the stability of drugs in formulations that may involve high temperatures, low pH, or other challenging conditions, thus extending the shelf life of pharmaceutical products. By studying their structure and function, researchers can gain insights into protein stability and adaptation strategies. This knowledge can be applied to engineer enzymes for various pharmaceutical purposes, including enhancing stability and catalytic activity.

Tannin acyl hydrolases find valuable applications within the pharmaceutical industry. These enzymes can be sourced from fungi such as *Penicillium variotii* and *Penicillium aspergillus*. To produce tannase, solid-state fermentation requires a substrate of 0.1–10% tannic acid. Following production, the enzyme is subjected to downstream processing and purification, often utilizing ion-exchange or gel-filtration chromatography methods (Gomes & Steiner, 2004). Pullulanase is instrumental in maltose production and also plays a role in cyclodextrin synthesis which serve as hosts for essential pharmaceutical compounds. The debranching attributes of pullulanases are employed in starch digestion (Kang et al., 2011). Esterases are employed in fine chemical and pharmaceutical production. They have been integral in the separation of racemic alcoholic and acid mixtures, yielding optically pure compounds. Many hyper/thermophilic esterases are affiliated with archaeal genera such as Pyrococcus, Sulfolobus and Pyrobaculum (Shao et al., 2014). Amidases contribute to the conversion of amides into carboxylic acids and ammonia, crucial building blocks in fine chemical and pharmaceutical synthesis. Thermophilic archaea like Sulfolobus (Mueller et al., 2006) and Pyrococcus (Fu et al., 2014) have been found to house these enzymes. Nitrilases are pivotal in generating pharmaceutical drug ingredients and intermediates, catalyzing the synthesis of compounds such as antidepressants, anti-inflammatory drugs, anticholesterol drugs, analgesics, antitumor agents, antiallergic drugs, antibiotics, and more (Gong et al., 2012).

Subtilase, derived from the thermophilic bacterium *Coprothermobacter proteolyticus*, belongs to a protein family widely utilized in pharmaceutical industries. Expression of the thermostable subtilase enzyme in *Escherichia coli* is facilitated in the presence of organic solvents and detergents, even at high temperatures of up to 80 °C (Toplak et al., 2013). Additionally, a thermostable nucleoside phosphorylase from hyperthermophilic *Aeropyrum pernix* K1 has been harnessed for synthesizing nucleoside analogs, proving useful in antiviral therapies as an alternative to chemical synthesis (Zhu et al., 2013). Thermostable lipases are of paramount importance in the pharmaceutical sector for processes like hydrolysis, transesterification, grease esterification, interesterification, and organic biosynthesis. A reported instance is the cloning and characterization of a thermostable lipase from Geobacillus sp. EPT9 in deep-sea hydrothermal fields. This recombinant lipase maintains 44% residual activity after 1 of incubation at 80 °C, and a monoacylglycerol lipase isolated from thermophilic bacterium Geobacillus sp. in the Arctic's deep-sea hydrothermal vent site (Tang et al., 2019). Thermophilic DNA polymerases are used in the polymerase chain reaction (PCR), a widely used molecular biology technique for amplifying DNA sequences. These enzymes enable DNA amplification at high temperatures, facilitating various genetic and diagnostic applications (Perzborn et al., 2010).

It's important to note that while extremozymes hold significant potential in the pharmaceutical industry, their practical implementation might require overcoming challenges related to enzyme stability, specificity, and regulatory approval. Nonetheless, ongoing research and advancements in enzyme engineering are paving the way for their increased use in pharmaceutical applications.

### 3.4.5 Other Potential Applications

Extremophilic enzymes, including proteases, lipases, cellulases, and amylases, have gained prominence as commercial enzymes widely employed in various industries, particularly in detergents (Chang et al., 2004). The use of enzymes in detergent production constitutes a substantial market share, accounting for around 40% of the global enzyme production, encompassing both large-scale manufacturing and high-value-added products (Margesin et al., 2008). Cellulases and amylases play pivotal roles in processes such as desizing, biofinishing, and stone washing, contributing to the removal of surface fibrils, pills, and the enhancement of fabric texture (Collins et al., 2007).

The remarkable diversity and exceptional characteristics of extremozymes, including their reproducibility, high efficiency, and economic viability, have spurred their integration into various biotechnological applications across diverse industrial processes (Gurung et al., 2013). More recently, radiation-resistant extremophiles have been explored for their biological activities, including potential applications in anticancer drugs, antioxidants, and sunscreens (Singh & Gabani, 2011). Stress conditions trigger gene modulation for enzyme synthesis, yielding important biomedical chemical compounds and potentially influencing bacterial growth inhibition mechanisms (Ameer et al., 2020). The alteration of physiological and metabolic traits under stress conditions is particularly evident in salt-tolerant bacteria and actinomycetes, which produce various pharmacologically relevant metabolites (Singh & Gabani, 2011).

Extremophilic enzymes, such as cellulases and laccases, have found utility in the textile industry for processes like biofinishing and color removal. Cellulases facilitate textile softening, while laccases contribute to color removal without the use of harsh chemicals. Their stability under extreme conditions makes them indispensable tools in textile processing (Siddiqui & Cavicchioli, 2006). Given that textile effluents often exhibit high salt levels, alkalinity, and elevated temperatures, extremophilic microbes hold significant biological importance due to their capacity to thrive in such challenging environments (Eslami et al., 2016). Numerous studies have demonstrated the potential of extremozymes from diverse extremophiles in dye degradation. Alkali, halo, and thermo-tolerant strains have displayed remarkable dye removal capabilities under acidic conditions (Kaushik et al., 2014). Similarly, extremophilic cyanobacteria exhibited an 80% removal efficiency for textile dye in alkaline industrial wastewater (Bela & Malliga, 2015). Furthermore, thermophilic Anoxybacillus spp. have been harnessed to successfully remove dyes from textile effluents (Goh et al., 2013). Employing microbial consortiums through bioaugmentation and biostimulation approaches has resulted in impressive decolorization rates of textile effluents, reaching up to 92% (Lade et al., 2015).

## 3.5 Challenges and Future Perspectives

Bioprospecting of extremophiles for industrial enzymes is a promising avenue for scientific advancement and technological innovation, yet it gives rise to a range of ethical concerns. While the potential benefits are significant, including novel enzyme discovery and industrial applications, the conservation of biodiversity must be a paramount consideration. Collecting these unique microorganisms from their natural habitats could disturb delicate ecosystems and result in biodiversity loss. Equitable access and benefit sharing are crucial, particularly when extremophiles are sourced from regions rich in biodiversity. Respect for indigenous knowledge and engagement with local communities are imperative, ensuring that traditional wisdom and perspectives are acknowledged and integrated. Additionally, ethical questions surround the ownership of genetic information and intellectual property rights stemming from these discoveries. Creating balance between short-term financial gains and long-term ecological and scientific benefits is essential. Regulatory oversight, transparency, and responsible collaboration among researchers, industry, communities, and governments are key components in navigating the ethical landscape of extremophile bioprospecting. Ultimately, a holistic approach that values biodiversity, respects cultural knowledge, and upholds equitable practices is necessary to ensure that bioprospecting contributes positively to both science and society.

In the exploration of extremophiles for industrial enzyme applications, ethical and environmental considerations play a pivotal role. Ethically, the preservation of biodiversity and indigenous rights must be paramount. The potential impact of extracting extremophiles from their natural habitats on local ecosystems and communities should be carefully evaluated, with a focus on sustainable practices and equitable benefit-sharing. Moreover, the responsible use of genetic information and fair patenting practices should be upheld, acknowledging the collaborative nature of scientific discovery. Environmentally, the processes involved in bioprospecting, enzyme production, and potential release of genetically modified organisms must be conducted with vigilance. Striking a balance between harnessing extremophile resources for innovation and safeguarding the delicate ecosystems they inhabit is pivotal to ensure the responsible advancement of biotechnology.

## 3.6 Conclusion

The bioprospecting of extremophiles for industrial enzymes holds exciting prospects for multiple fields. Advances in genomic analysis and synthetic biology are expected to unlock the potential of extremophile enzymes, enabling tailored modifications that enhance their properties for specific applications. The integration of computational tools and machine learning will accelerate the discovery process by predicting enzyme functionalities and interactions. Collaborations between researchers, indigenous communities, and industry stakeholders will drive ethical and sustainable practices, ensuring equitable benefit-sharing and the preservation of biodiversity. As extremophile-based biotechnology gains traction, innovative enzyme applications in sectors such as renewable energy, bioremediation, and personalized medicine are anticipated. Moreover, the integration of extremophiles into biotechnological processes has the potential to reduce resource consumption and waste, contributing to a more sustainable and environmentally conscious future. As we harness the unique capabilities of extremophiles and their enzymes, interdisciplinary efforts will pave the way for novel solutions to global challenges, making their mark on industries and research landscapes alike.

## References


Abe, F., & Horikoshi, K. (2001). The biotechnological potential of piezophiles. *Trends in Biotechnology, 19*, 102–108.

Alain, K., Callac, N., Guégan, M., Lesongeur, F., Crassous, P., Cambon-Bonavita, M. A., Querellou, J., & Prieur, D. (2009). Nautilia abyssi sp. nov., a thermophilic, chemolithoautotrophic, sulfur-reducing bacterium isolated from an East Pacific Rise hydrothermal vent. *International Journal of Systematic and Evolutionary Microbiology, 59*, 1310–1315.

Ameer, K., Chirom, A., & Paul, A. (2020). Production and purification of anti-tubercular and anticancer protein from Staphylococcus hominis under mild stress condition of Mentha piperita L. *Journal of Pharmaceutical and Biomedical Analysis, 182*. https://doi.org/10.1016/j.jpba.2020.113136

Aono, R., & Horikoshi, K. (1983). Hydrocarbon-degrading Alcaligenes sp. isolated from deep-sea sediment. *Applied and Environmental Microbiology, 45*(5), 1602–1608.

Arun, A. B., Chen, W. M., Lai, W. A., Chou, J. H., Shen, F. T., Rekha, P. D., & Young, C. C. (2009). Lutaonella thermophila gen. nov., sp. nov., a moderately thermophilic member of the family Flavobacteriaceae isolated from a coastal hot spring. *International Journal of Systematic and Evolutionary Microbiology, 8*, 2069–2073.

Averhoff, B., & Müller, V. (2010). Exploring research frontiers in microbiology: Recent advances in halophilic and thermophilic extremophiles. *Research in Microbiology, 161*, 506–514.

Aygan, A., Arikan, B., Korkmaz, H., Dincer, S., & Colak, O. (2008). Highly thermostable and alkaline α-amylase from a halotolerant-alkaliphilic Bacillus sp. AB68. *Brazilian Journal of Microbiology, 39*(3), 547–553. https://doi.org/10.1590/S1517-83822008000300027

Baborova, P., Moder, M., Baldrian, P., Cajthamlva, K., & Cajthaml, T. (2006). Purification of a new manganese peroxidase of the white-rot fungus Irpex lacteus and degradation of polycyclic aromatic hydrocarbons by the enzyme. *Research in Microbiology, 157*, 248.

Bajpai, P. (2016). Extremophiles-based enzymes for food applications. *Food and Bioprocess Technology, 9*(2), 191–203.

Baker-Austin, C., & Dopson, M. (2007). Life in acid: pH homeostasis in acidophiles. *Trends in Microbiology, 15*, 165–171.

Baliga, N. S., Bonneau, R., Facciotti, M. T., Pan, M., Glusman, G., Deutsch, E. W., et al. (2004). Genome sequence of Haloarcula marismortui: A halophilic archaeon from the Dead Sea. *Genome Research, 14*(11), 2221–2234.

Bartlett, D. H. (2002). Pressure effects on in vivo microbial processes. *Biochimica et Biophysica Acta (BBA): Proteins and Proteomics, 1595*(1–2), 367–381.

Barcelos, M. C., Ramos, C. L., Kuddus, M., Rodriguez-Couto, S., Srivastava, N, Ramteke, P. W., Mishra, P. K., & Molina, G. (2020). Enzymatic potential for the valorization of agro-industrial by-products. *Biotechnology Letters*, 1–29. https://doi.org/10.1007/s10529-020-02957-3

Battista, J. R., & Cox, M. M. (2005). Protein-membrane interactions: The assembly of RecA protein onto lipid bilayers is a radiation-resistant process. *Molecular Microbiology, 55*(2), 511–521.

Basu, S., & Sen, S. (2009). Turning a mesophilic protein into a thermophilic one: A computational approach based on 3D structural features. *Journal of Chemical Information and Modeling, 49*, 1741–1750.

Bela, R. B., & Malliga, P. (2015). Treatment of textile dye effluent using marine cyanobacterium Lyngbya sp. with different agrowastes and its effect on the growth of cyanobacterium. *Journal of Environmental Biology, 36*(3), 623–626.

Bertoldo, C. & Antranikian, G. (2002). Starch-hydrolyzing enzymes from thermophilic archaea and bacteria. *Current Opinion in Chemical Biology, 6*, 151–160.

Bhaskar, N., Mahendran, R., & Renganathan, S. (2007). Optimization of alkaline protease production by Bacillus sphaericus. *Journal of Agricultural and Food Chemistry, 55*(10), 4080–4086.

Bhatt, P., Sharma, A., Gangola, S., Khati, P., Kumar, G., & Srivastava, A. (2016). Novel pathway of cypermethrin biodegradation in a Bacillus sp. strain SG2 isolated from cypermethrin-contaminated agriculture field. *3 Biotech, 6*, 1.

Bhatt, P., Gangola, S., Chaudhary, P., Khati, P., Kumar, G., Sharma, A., & Srivastava, A. (2019). Pesticide induced up-regulation of esterase and aldehyde dehydrogenase in indigenous Bacillus spp. *Bioremediation Journal, 23*, 42.

Boonyaratanakornkit, B. B., Park, C. B., & Clark, D. S. (2002). Pressure effects on intra- and intermolecular interactions within proteins. *Biochimica et Biophysica Acta, 1595*, 235–249.

Borges, D. G., Junior, A. B., Farinas, C. S., Giordano, R. L. C., & Tardioli, P. W. (2014). Enhanced saccharification of sugarcane bagasse using soluble cellulase supplemented with immobilized β-Glucosidase. *Bioresource Technology, 167*, 206–213. https://doi.org/10.1016/j.biortech.2014.06.021

Brasen, C., Urbanke, C., & Schonheit, P. (2005). A novel octameric AMP-forming acetyl-CoA synthetase from the hyperthermophilic crenarchaeon Pyrobaculum aerophilum. *FEBS Letters, 579*, 477–482.

Brock, T. D. (1978). *Thermophilic microorganisms and life at high temperatures*. Springer-Verlag.

Bustard, M. T., Burgess, J. G., Meeyoo, V., & Wright, P. C. (2000). Novel opportunities for marine hyperthermophiles in emerging biotechnology and engineering industries. *Journal of Chemical Technology and Biotechnology, 75*, 1095–1109.

Bowman, J. P., McCammon, S. A., Brown, M. V., Nichols, D. S., & McMeekin, T. A. (1997). Diversity and association of psychrophilic bacteria in Antarctic sea ice. *Applied and Environmental Microbiology, 63*, 3068–3078.

Cannio, R., Di Prizito, N., Rossi, M., & Morana, A. (2004). A xylan-degrading strain of Sulfolobus solfataricus: Isolation and characterization of the xylanase activity. *Extremophiles, 8*, 117–124.

Cavicchioli, R., Thomas, T. (2000). Extremophiles in: encyclopedia of microbiology vol 2, 2nd edn. Academic, London, pp 317–337.

Cavicchioli, R., Siddiqui, K. S., Andrews, D., & Sowers, K. R. (2002). Low-temperature extremophiles and their applications. *Current Opinion in Biotechnology, 13*, 253–261.

Chandran, J., Amma, K. P. P., Menon, N., Purushothaman, J., & Nisha, P. (2012). Effect of enzyme assisted extraction on quality and yield of volatile oil from black pepper and cardamom. *Food Science and Biotechnology, 21*(4), 1611–1617.

Chang, P., Tsai, W. S., Tsai, C. L., & Tseng, M. J. (2004). Cloning and characterization of two thermostable xylanases from an alkaliphilic Bacillus firmus. *Biochemical and Biophysical Research Communications, 319*, 1017–1025.

Chaplin, M. (2015). *Applications of proteases in the food industry*. Retrieved from http://www1.lsbu.ac.uk/water/enztech/proteases.html

Chaturvedi, P., Prabahar, V., Manorama, R., Pindi, P. K., Bhadra, B., Begum, Z., & Shivaji, S. (2008). Exiguobacterium soli sp. nov., a psychrophilic bacterium from the McMurdo Dry Valleys, Antarctica. *International Journal of Systematic and Evolutionary Microbiology, 58*, 2447–2453.

Cheng, F. S., Sheng, J. P., Dong, R. B., Men, Y. J., Gan, L., & Shen, L. (2012). Novel xylanase from a Holstein cattle rumen metagenomic library and its application in xylo-oligosaccharide and ferulic acid production from wheat straw. *Journal of Agricultural and Food Chemistry, 60*(51), 12516–12524.

Christopher, L. P., Zambare, V. P., Zambare, A., Kumar, H., & Malek, L. (2015). A thermo-alkaline lipase from a new thermophile Geobacillus thermodenitrificans AV-5 with potential application in biodiesel production. *Journal of Chemical Technology & Biotechnology, 90*(11), 2007–2016.

Collins, T., D'Amico, S., Marx, J. C., Feller, G., & Gerday, C. (2007). Extremozymes industrial applications. *Biotechnology Advances, 25*(6), 547–591.

Cordone, L., Ferrand, M., Vitrano, E., & Zaccai, G. (1999). Harmonic behavior of trehalose-coated carbon-monoxymyoglobin at high temperature. *Biophysical Journal, 76*, 1043–1047.

Coronado, M., Vargas, C., Hofemeister, J., Ventosa, A., & Nieto, J. J. (2000). Production and biochemical characterization of an alpha-amylase from the moderate halophile Halomonas meridiana. *FEMS Microbiology Letters, 183*(1), 67–71.

Cox, M. M., & Battista, J. R. (2005). Deinococcus radiodurans—The consummate survivor. *Nature Reviews Microbiology, 3*(11), 882–892.

da Fonseca, F. S., Angolini, C. F., Arruda, M. A., Junior, C. A., Santos, C. A., Saraiva, A. M., et al. (2015). Identification of oxidoreductases from the petroleum Bacillus safensis strain. *Biotechnology Reports, 8*, 152–159. https://doi.org/10.1016/j.btre.2015.09.001

Dadachova, E., & Casadevall, A. (2007). Cryptococcus neoformans: A loveable yeast? *FEMS Yeast Research, 7*(3), 464–468.

Daly, M. J. (2009). A new perspective on radiation resistance based on Deinococcus radiodurans. *Nature Reviews Microbiology, 7*(3), 237–245.

Daly, M. J. (2012). Death by protein damage in irradiated cells. *DNA Repair, 11*(1), 12–21.

Daniel, R. M. (1996). The upper limits of enzyme thermostability. *Enzyme and Microbial Technology, 19*, 74–79.

Datta, S., Holmes, B., Park, J., Chen, Z., Dibble, D. C., Hadi, M., et al. (2010). Ionic liquid tolerant hyperthermophilic cellulases for biomass pretreatment and hydrolysis. *Green Chemistry, 12*, 338–345.

de Camargo, B. R., Claassens, N. J., Quirino, B. F., Noronha, E. F., & Kengen, S. W. (2018). Heterologous expression and characterization of a putative glycoside hydrolase family 43 arabinofuranosidase from Clostridium thermocellum B8. *Enzyme and Microbial Technology, 109*, 74–83.

Declerck, N., Machius, M., Joyet, P., Wiegand, G., & Huber, R. (2003). Hyper thermostabilization of Bacillus licheniformis α-amylase and modulation of its stability over a 50°C temperature range. *Protein Engineering, 16*(4), 287–293.

Del Campo, J. S. M., Chun, Y., Kim, J. E., Patiño, R., & Zhang, Y. H. P. (2013). Discovery and characterization of a novel ATP/polyphosphate xylulokinase from a hyperthermophilic bacterium Thermotoga maritima. *Journal of Industrial Microbiology & Biotechnology, 40*(7), 661–669.

Dharmsthiti, S., & Luchai, S. (1999). Production, purification and characterization of thermophilic lipase from Bacillus sp. THL027. *FEMS Microbiology Letters, 179*(2), 241–246.

Dhiman, T. R., Zaman, M. S., Gimenez, R. R., Walters, J. L., & Treacher, R. (2002). Performance of dairy cows fed forage treated with fibrolytic enzymes prior to feeding. *Animal Feed Science and Technology, 101*(1–4), 115–125.

Di Donato, P., Buono, A., Poli, A., Finore, I., Abbamondi, G. R., Nicolaus, B., & Lama, L. (2019). Exploring marine environments for the identification of extremophiles and their enzymes for sustainable and green bioprocesses. *Sustainability, 11*, 149. https://doi.org/10.3390/su11010149

Dopson, M., Baker-Austin, C., Hind, A., Bowman, J. P., & Bond, P. L. (2004). Characterization of Ferroplasma Isolates and Ferroplasma acidarmanus sp. nov., extreme acidophiles from acid mine drainage and industrial bioleaching environments. *Applied and Environmental Microbiology, 70*, 2079–2088.

Duckworth, A. W., Grant, W. D., Jones, B. E., & Van Steenbergen, R. (1996). Phylogenetic diversity of soda lake alkaliphiles. *FEMS Microbiology Ecology, 19*, 181–191.

Duffner, F., Bertoldo, C., Andersen, J. T., Wagner, K., & Antranikian, G. (2000). A new thermoactive pullulanase from Desulfurococcus mucosus: Cloning, sequencing, purification, and characterization of the recombinant enzyme after expression in Bacillus subtilis. *Journal of Bacteriology, 182*(22), 6331–6338.

Dumorne, K., Cordova, D. C., Astorga-Elo, M., & Renganathan, P. (2017). Extremozymes: A potential source for industrial applications. *Journal of Microbiology and Biotechnology 27*, 649.

Engel, A. S., Johnson, L. R., & Porter, M. L. (2013). Arsenite oxidase gene diversity among Chloroflexi and Proteobacteria from El Tatio geyser field, Chile. *FEMS Microbiology Ecology, 83*(3), 745–756. https://doi.org/10.1111/1574-6941.12030

Eslami, M., Amoozegar, M. A., & Asad, S. (2016). Isolation, cloning and characterization of an azoreductase from the halophilic bacterium Halomonas elongata. *International Journal of Biological Macromolecules, 85*, 111–116. https://doi.org/10.1016/j.ijbiomac.2015.12.065

Fang, J., Zhang, L., & Bazylinski, D. A. (2010). Deep-sea piezosphere and piezophiles: Geomicrobiology and biogeochemistry. *Trends in Microbiology, 18*, 413–422.

Fardeau, M. L., Barsotti, V., Cayol, J. L., Guasco, S., Michotey, V., Joseph, M., Bonin, P., & Ollivier, B. (2010). Caldinitratiruptor microaerophilus, gen. nov., sp. nov. isolated from a French hot spring (Chaudes-Aigues, Massif Central): A novel cultivated facultative microaerophilic anaerobic thermophile pertaining to the symbiobacterium branch within the Firmicutes. *Extremophiles, 14*, 241–247.

Febriani, R., Ulya, M., Oesman, F., & Iqbalsyah, T. M. (2019). Low molecular weight alkaline thermostable α-amylase from Geobacillus sp. nov. *Heliyon, 5*(7), e02171.

Feller, G. (2003). Molecular adaptations to cold in psychrophilic enzymes. *Cellular and Molecular Life Sciences, 60*, 648–662. https://doi.org/10.1007/s00018-003-2155-3

Feller, G., & Gerday, C. (2003). Psychrophilic enzymes: Hot topics in cold adaptation. *Nature Reviews Microbiology, 1*(3), 200–208.

Fitter, J., Herrmann, R., Dencher, N. A., Blume, A., & Hauss, T. (2001). Activity and stability of a thermostable α-amylase compared to its mesophilic homologue: Mechanisms of thermal adaptation. *Biochemistry, 40*(35), 10723–10731.

Florczak, T., Daroch, M., Wilkinson, M. C., Białkowska, A., Bates, A. D., Turkiewicz, M., & Iwanejko, L. A. (2013). Purification, characterisation and expression in Saccharomyces cerevisiae of LipG7 an enantioselective, cold-adapted lipase from the Antarctic filamentous fungus Geomyces sp. P7 with unusual thermostability characteristics. *Enzyme and Microbial Technology, 53*(1), 18–24. https://doi.org/10.1016/j.enzmictec.2013.03.021

Friedmann, E. I. (1993). *Antarctic microbiology*. Wiley-Liss.

Fu, L., et al. (2014). Purification and characterization of a thermostable aliphatic amidase from the hyperthermophilic archaeon Pyrococcus yayanosii CH1. *Extremophiles, 18*(2), 429–440.

Fukushima, T., Mizuki, T., Echigo, A., Inoue, A., & Usami, R. (2005). Organic solvent tolerance of halophilic α-amylase from a haloarchaeon, Haloarcula sp. strain S-1. *Extremophiles, 9*, 85–89.

Fusi, P., Grisa, M., Mombelli, E., Consonni, R., Tortora, P., & Vanoni, M. (1995). Expression of a synthetic gene encoding P2 ribonuclease from the extreme thermoacidophilic archaebacterium Sulfolobus solfataricus in mesophylic hosts. *Gene, 154*, 99–103.

Gänzle, M. G. (2015). Enzymatic and bacterial conversions during sourdough fermentation. *Food Microbiology, 37*, 2–10.

Gaur, R., Tiwari, S., & Goyal, D. (2015). Thermophilic and alkaliphilic cellulases: Versatile catalysts for biomass deconstruction. *Critical Reviews in Biotechnology, 35*(1), 69–86.

Gerday, C., Aittaleb, M., Arpigny, J. L., Baise, E., Chessa, J.-P., Garsoux, G., Petrescu, I., & Feller, G. (1997). Psychrophilic enzymes: A thermodynamic challenge. *Biochimica et Biophysica Acta, Protein Structure and Molecular Enzymology, 1342*, 119–131.

Georlette, D., Blaise, V., Collins, T., D'Amico, S., Gratia, E., Hoyoux, A., et al. (2004). Some like it cold: Biocatalysis at low temperatures. *FEMS Microbiology Reviews, 28*, 25–42.

Ghorbel, R. E., Maktouf, S., Massoud, E. B., Bejar, S., & Chaabouni, S. E. (2009). New thermostable amylase from Bacillus cohnii US147 using solid-state fermentation: Purification and characterization. *Applied Biochemistry and Biotechnology, 157*(3), 468–480.

Goh, K. M., et al. (2013). Recent discoveries and applications of Anoxybacillus. *Applied Microbiology and Biotechnology, 97*(4), 1475–1488. https://doi.org/10.1007/s00253-012-4663-2

Golyshina, O., & Timmis, K. N. (2005). Ferroplasma and relatives, recently discovered cell wall-lacking archaea making a living in extremely acid, heavy metal-rich environments. *Environmental Microbiology, 7*, 1277–1288.

Gomes, J., & Steiner, W. (2004). Extremophiles and Extremozymes. *Food Technology and Biotechnology, 42*(4), 223–235.

Gong, J. S., et al. (2012). Nitrilases in nitrile biocatalysis: Recent progress and forthcoming research. *Microbial Cell Factories, 11*, 142.

Grant, W. D., Mwatha, W. E., & Jones, B. E. (1990). Alkaliphiles: Ecology, diversity and applications. *FEMS Microbiology Reviews, 75*, 255–270.

Gupta, R., & Gigras, P. (2014). Microbial α-amylases: A biotechnological perspective. *Process Biochemistry, 38*(11), 1599–1616.

Gupta, R., Beg, Q. K., & Lorenz, P. (2002). Bacterial alkaline proteases: Molecular approaches and industrial applications. *Applied Microbiology and Biotechnology, 59*(1), 15–32.

Gurung, N., Ray, S., Bose, S., & Rai, V. (2013). A broader view: Microbial enzymes and their relevance in industries, medicine, and beyond. *Biomed Research International, 2013*, 329121.

Gyure, R. A., Konopka, A., Brooks, A., & Doemel, W. (1987). Algal and bacterial activities in acidic (pH 3) strip mine lakes. *Applied and Environmental Microbiology, 53*, 2069–2076.

Hasan, F., Shah, A. A., Hameed, A., & Ahmed, S. (2006). Industrial applications of microbial lipases. *Enzyme and Microbial Technology, 39*(2), 235–251.

Hauenstein, S., Zhang, C. M., Hou, Y. M., & Perona, J. J. (2004). Shape-selective RNA recognition by cysteinyl-tRNA synthetase. *Nature Structural & Molecular Biology, 11*, 1134–1141.

Hoekman, S. K. (2009). Biofuels in the US: Challenges and opportunities. *Renewable Energy, 34*(1), 14–22.

Huang, Y., Krauss, G., Cottaz, H., Driguez, H., & Lipps, G. (2005). A highly acid-stable and thermostable endo-β-glucanase from the thermoacidophilic archaeon Sulfolobus solfataricus. *Biochemical Journal, 385*, 581–588.

Irwin, J. A. (2020). Chapter 6—Overview of extremophiles and their food and medical applications. In R. Salwan & V. Sharma (Eds.), *Physiological and biotechnological aspects of extremophiles* (pp. 65–87). Academic. https://doi.org/10.1016/B978-0-12-818322-9.00006-X

Ito, M., & Guffanti, A. A. (2002). Alkaline pH homeostasis in bacteria: New insights. *Biochimica et Biophysica Acta (BBA)-Biomembranes, 1559*(1), 2–9.

Jackson, B. R., Noble, C., Lavesa-Curto, M., Bond, P. L., & Bowater, R. P. (2007). Characterization of an ATP-dependent DNA ligase from the acidophilic archaeon "Ferroplasma acidarmanus" Fer1. *Extremophiles, 11*, 315–327.

Jaeger, K. E., & Eggert, T. (2002). Lipases for biotechnology. *Current Opinion in Biotechnology, 13*(4), 390–397.

Jaenicke, R. (1981). Enzymes under extreme of physical conditions. *Annual Review of Biochemistry, 75*, 403–433.

Jaenicke, R., Schuring, H., Beaucamp, N., & Ostendorp, R. (1996). Structure and stability of hyperstable proteins: Glycolytic enzymes from hyperthermophilic bacterium Thermotoga maritima. *Advances in Protein Chemistry, 48*, 181–269.

Johnson, D. B., & Hallberg, K. B. (2008). Carbon, iron and sulfur metabolism in acidophilic microorganisms. *Advances in Microbial Physiology, 54*, 201–255.

Kaledhonkar, M. J., Naqvi, S., & Gowrishankar, J. (2016). PCR fidelity of Pfu DNA polymerase and other thermostable DNA polymerases. *Nucleic Acids Research, 44*(4), e31.

Kang, J., et al. (2011). Molecular cloning and biochemical characterization of a heat-stable type I pullulanase from Thermotoga neapolitana. *Enzyme Microbial Technology, 48*(3), 260–266.

Karbalaei-Heidari, H. R., Ziaee, A. A., & Amoozegar, M. A. (2007). Purification and biochemical characterization of a protease secreted by the Salinivibrio sp. strain AF-2004 and its behavior in organic solvents. *Extremophiles, 11*, 237–243.

Kaushik, P., et al. (2014). Biosorption of textile dye by Aspergillus lentulus pellets: Process optimization and cyclic removal in aerated bioreactor. *Water, Air, & Soil Pollution, 225*(6). https://doi.org/10.1007/s11270-014-1978-x

Khelifi, N., Ben Romdhane, E., Hedi, A., Postec, A., Fardeau, M. L., Hamdi, M., Tholozan, J. L., Ollivier, B., & Hirschler-Réa, A. (2010). Characterization of Microaerobacter geothermalis gen. nov., sp. nov., a novel microaerophilic, nitrate- and nitrite-reducing thermophilic bacterium isolated from a terrestrial hot spring in Tunisia. *Extremophiles, 14*, 297–304.

Kim, K. H., Choi, I. S., Kim, H. M., Wi, S. G., & Bae, H. J. (2014). Bioethanol production from the nutrient stress-induced microalga Chlorella vulgaris by enzymatic hydrolysis and immobilized yeast fermentation. *Bioresource Technology, 153*, 47–54.

Klibanov, A. M., Tu, T. M., & Scott, K. P. (1983). Peroxidase-catalyzed removal of phenols from coal-conversion wastewaters. *Science, 221*, 259.

Kuddus, M., & Singh, A. (2012). Thermophilic amylase: Current status and recent developments. *Recent Patents on Biotechnology, 6*(2), 154–161.

Kumar, A., & Goyal, A. (2019). Applications of acidophilic microbes for the bioleaching of metals. *Transactions of the Indian National Academy of Engineering, 4*(1), 1–14.

Kumar, M., Vladimir, L., de Sistro Materano, A., & Ilzins, O. A. (2007). A halotolerant and thermotolerant Bacillus sp. degrades hydrocarbons and produces tension active emulsifying agent. *World Journal of Microbiology and Biotechnology, 23*, 211–220.

Kumar, S., Karan, R., Kapoor, S., Singh, S. P., & Khare, S. K. (2012). Screening and isolation of halophilic bacteria producing industrially important enzymes. *Brazilian Journal of Microbiology, 43*(4), 1595–1603.

Kumar M, Yadav A. N., Saxena R, Paul D, Tomar R. S. (2021). Biodiversity of pesticides degrading microbial communities and their environmental impact. *Biocatalysis and Agricultural Biotechnology, 31*, 101883.

Koyama, N. (1999). Presence of Na+-stimulated P-type ATPase in the membrane of a facultatively anaerobic alkaliphile, Exiguobacterium aurantiacum. *Current Microbiology, 39*, 27–30.

Lade, H., Govindwar, S., & Paul, D. (2015). Mineralization and detoxification of the carcinogenic azo dye Congo red and real textile effluent by a polyurethane foam immobilized microbial consortium in an upflow column bioreactor. *International Journal of Environmental Research and Public Health, 12*(6), 6894–6918. https://doi.org/10.3390/ijerph120606894

Lauro, F. M., & Bartlett, D. H. (2008). Prokaryotic lifestyles in deep sea habitats. *Extremophiles, 12*(1), 15–25.

Leduc, L. G., & Ferroni, G. D. (1994). The chemolithotrophic bacterium Thiobacillus ferrooxidans. *FEMS Microbiology Reviews, 14*, 103–120.

Lee, S. G., Ro, H. S., Hong, S. P., Kim, E. H., & Sung, M. H. (1996). Production of L-DOPA by thermostable tyrosine phenol-lyase of a thermophilic Symbiobacterium species overexpressed in recombinant Escherichia coli. *Journal of Microbiology and Biotechnology, 6*, 98.

Leigh, J. A., & Wolfe, R. S. (1983). Acetogenium kivui gen. nov., sp. nov., a thermophilic acetogenic bacterium. *International Journal of Systematic Bacteriology, 33*, 886–889.

Litchfield, C. D. (2011). Potential for industrial products from the halophilic Archaea. *Journal of Industrial Microbiology & Biotechnology, 38*, 1635–1647.

Lopez-Archilla, A. I., Marin, I., & Amils, R. (1995). Microbial ecology of an acidic river: Biotechnological applications. In T. Vargas, C. A. Jerez, J. V. Wiertz, & H. Toledo (Eds.), *Biohydrometallurgical processing II* (pp. 63–74). University of Chile.

Luque, R., Herrero-Davila, L., Campelo, J. M., Clark, J. H., Hidalgo, J. M., Luna, D., et al. (2008). Biofuels: A technological perspective. *Energy & Environmental Science, 1*(5), 542–564.

Makarova, K. S., et al. (2007). Deinococcus geothermalis: The pool of extreme radiation resistance genes shrinks. *PLoS One, 2*(9), e955.

Madigan, M. T., & Marrs, B. L. (1997). Extremophiles. *Scientific American, 276*, 82–87.

Margesin, R., Schinner, F., Marx, J. C., & Gerday, C. (Eds.). (2008). *Psychrophiles: From biodiversity to biotechnology*. Springer.

Margesin, R., & Fell, J. W. (2008). Mrakiella cryoconiti gen. nov., sp. nov., a psychrophilic, anamorphic, basidiomycetous yeast from alpine and Arctic habitats. *International Journal of Systematic and Evolutionary Microbiology, 58*, 2977–2982.

Mayer, F., Küper, U., Meyer, C., Daxer, S., Müller, V., Rachel, R., & Huber, H. (2012). AMP-forming acetyl coenzyme A synthetase in the outermost membrane of the hyperthermophilic crenarchaeon Ignicoccus hospitalis. *Journal of Bacteriology, 194*, 1572–1581.

Mesbah, N. M., & Wiegel, J. (2008). Life at extreme limits. The anaerobic halophilic alkalithermophiles. *Annals of the New York Academy of Sciences, 1125*, 44–57.

Mesbah, N., Cook, G., & Wiegel, J. (2009). The halophilic alkalithermophile Natranaerobius thermophilus adapts to multiple environmental extremes using a large repertoire of Na+ (K+)/H+ antiporters. *Molecular Microbiology, 74*, 270–281.

Mohanty, A., Shilpa, S., & Meena, S. S. (2022). Chapter 4: Microbial adaptation to extreme temperatures: An overview of molecular mechanisms to industrial application. In N. K. Arora, S. Agnihotri, & J. Mishra (Eds.), *Extremozymes and their industrial applications* (pp. 115–139). Academic. https://doi.org/10.1016/B978-0-323-90274-8.00009-5

Mombelli, E., Shehi, E., Fusi, P., & Tortora, P. (2002). Exploring hyperthermophilic proteins under pressure: Theoretical aspects and experimental findings. *Biochimica et Biophysica Acta, 1595*, 392–396.

Mori, K., Yamaguchi, K., Sakiyama, Y., Urabe, T., & Suzuki, K. I. (2009). Caldisericum exile gen. nov., sp. nov., an anaerobic, thermophilic, filamentous bacterium of a novel bacterial phylum, Caldiseria phyl. nov., originally called candidate phylum OP5. *International Journal of Systematic and Evolutionary Microbiology*. https://doi.org/10.1099/ijs.0.010033-0

Mueller, P., et al. (2006). Cloning, overexpression, and characterization of a thermoactive nitrilase from the hyperthermophilic archaeon Pyrococcus abyssi. *Protein Expression and Purification, 47*(2), 672–681.

Naik, S. N., Goud, V. V., Rout, P. K., & Dalai, A. K. (2010). Production of first and second generation biofuels: A comprehensive review. *Renewable and Sustainable Energy Reviews, 14*(2), 578–597.

Orellana, R., Macaya, C., Bravo, G., Dorochesi, F., Cumsille, A., Valencia, R., et al. (2018). Living at the frontiers of life: Extremophiles in Chile and their potential for bioremediation. *Frontiers in Microbiology, 9*, 2309. https://doi.org/10.3389/fmicb.2018.02309. PMID: 30425685.

Oshima, T., & Moriya, T. (2008). A preliminary analysis of microbial and biochemical properties of high-temperature compost. *Annals of the New York Academy of Sciences, 1125*, 338–344.

Ogg, C., & Patel, B. K. (2009). Sporolituus thermophilus gen. nov., sp. nov., a citrate-fermenting, thermophilic, anaerobic bacterium from geothermal waters of the Great Artesian Basin of Australia. *International Journal of Systematic and Evolutionary Microbiology, 59*, 2848–2853.

Padan, E., Bibi, E., Ito, M., & Krulwich, T. A. (2005). Alkaline pH homeostasin bacteria: New insights. *Biochimica et Biophysica Acta, Biomembranes, 1717*, 67–88.

Panda, S. K., Jyoti, V., Bhadra, B., Nayak, K. C., Shivaji, S., Rainey, F. A., & Das, S. K. (2009). Thiomonas bhubaneswarensis sp. nov., a novel obligately mixotrophic, moderately thermophilic, thiosulfate oxidizing bacterium. *International Journal of Systematic and Evolutionary Microbiology, 59*, 2171 2175.

Peeples, T. L. (2014). Bioremediation using extremophiles. In *Microbial biodegradation and bioremediation*. Elsevier Inc.

Pérez-Pomares, F., Bautista, V., Ferrer, J., Pire, C., Marhuenda-Egea, F. C., & Bonete, M. J. (2003). Alpha-amylase activity from the halophilic archaeon Haloferax mediterranei. *Extremophiles, 7*(4), 299–306.

Perzborn, E., Strassburger, J., Wilmen, A., Pohlmann, J., & Roehrig, S. (2010). In vitro and in vivo studies of the novel antithrombotic agent BAY 59-7939—An oral, direct factor Xa inhibitor. *Journal of Thrombosis and Haemostasis, 3*(3), 514–521.

Pozdnyakova, N. N., Rodakiewicz-Nowak, J., Turkovskaya, O. V., & Haber, J. (2006). Oxidative degradation of polyaromatic hydrocarbons catalyzed by blue laccase from Pleurotus ostreatus DI in the presence of synthetic mediators. *Enzyme and Microbial Technology, 39*, 1242.

Poli, A., Romano, I., Cordella, P., Orlando, P., Nicolaus, B., & Ceschi Berrini, C. (2009). Anoxybacillus thermarum sp. nov., a novel thermophilic bacterium isolated from thermal mud in Euganean hot springs, Abano Terme, Italy. *Extremophiles, 13*, 867–874.

Prevost, S., Andre, S., & Remize, F. (2010). PCR detection of thermophilic spore-forming bacteria involved in canned food spoilage. *Current Microbiology, 61*, 525–533.

Raddadi, N., Cherif, A., Daffonchio, D., Mohamed, N., & Fava, F. (2015). Biotechnological applications of extremophiles, extremozymes and extremolytes. *Applied Microbiology and Biotechnology, 99*, 7907–7913.

Rainey F.A., Donnison A. M., Janssen P. H., Saul D., Rodrigo A., Bergquist P.L., Daniel R.M., Stackebrandt E., Morgan H. W. (1994). Description of Caldicellulosiruptor saccharolyticus gen. nov., sp. nov: an obligately anaerobic, extremely thermophilic, cellulolytic bacterium. *FEMS Microbiol Lett 120*, 263–266.

Ray, M. K., Kumar, G. S., Janiyani, K., Kannan, K., Jagtap, P., Basu, M. K., & Shivaji, S. (1998). Adaptation to low temperature and regulation of gene expression in Antarctic psychrotrophic bacteria. *Journal of Biosciences, 23*, 423–435.

Reed, C. J., Lewis, H., Trejo, E., Winston, V., & Evilia, C. (2013). Protein adaptations in archaeal extremophiles. *Archaea, 2013*, 373275.

Romano, I., Dipasquale, L., Orlando, P., Lama, L., d'Ippolito, G., Pascual, J., & Gambacorta, A. (2010). Thermoanaerobacterium thermostercus sp. nov., a new anaerobic thermophilic hydrogen-producing bacterium from buffalo-dung. *Extremophiles, 14*, 233–240.

Ruiz, D. M., & De Castro, R. E. (2007). Effect of organic solvents on the activity and stability of an extracellular protease secreted by the haloalkaliphilic archaeon Natrialba magadii. *Journal of Industrial Microbiology & Biotechnology, 34*, 111–115.

Russell, N. J. (1997). Psychrophilic bacteria—molecular adaptations of membrane lipids. *Comparative Biochemistry and Physiology. Part A, Physiology, 118*, 489–493.

Saito, A., Iwabuchi, T., & Harayama, S. (2000). A novel phenanthrene dioxygenase from Nocardioides sp. strain KP7: Expression in Escherichia coli. *Journal of Bacteriology, 182*, 2134.

Salwan, R., & Sharma, V. (2020). Chapter 1: Overview of extremophiles. In R. Salwan & V. Sharma (Eds.), *Physiological and biotechnological aspects of extremophiles* (pp. 3–11). Academic. https://doi.org/10.1016/B978-0-12-818322-9.00001-0

Sathya, T. A., Methew, A. J., & Khan, M. (2014). Cloning and Molecular modeling of pectin degrading Glycosyl Hydrolase of family 28 from soil metagenomic library. *Molecular Biology Reports, 41*, 2645–2456.

Saxena, A. K., Padaria, J. C., Gurjar, G. T., Yadav, A. N., Lone, S. A., & Tripathi, M. S. (2020). RMV: Insecticidal formulation of novel strain of Bacillus thuringiensis AK 47. In Indian Patent 340541.

Scoma, A., & Boon, N. (2016). Marine extremophiles as a source of bioactive molecules. *Marine Genomics, 27*, 51–57.

Sellek, G. A., & Chaudhuri, J. B. (1999). Biocatalysis in organic media using enzymes from extremophiles. *Enzyme and Microbial Technology, 25*, 471–482.

Shafiei, M., Ziaee, A. A., & Amoozegar, M. A. (2011). Purification and characterization of an organic-solvent-tolerant halophilic α-amylase from the moderately halophilic Nesterenkonia sp. strain F. *Journal of Industrial Microbiology & Biotechnology, 38*, 275–281.

Shao, H., Xu, L., & Yan, Y. (2014). Biochemical characterization of a carboxylesterase from the archaeon pyrobaculum sp. 1860 and a rational explanation of its substrate specificity and thermostability. *International Journal of Molecular Sciences, 15*(9), 16885–16910.

Sharma, D., Sharma, B., & Shukla, A. K. (2011). Biotechnological approach of microbial lipase: A review. *Biotechnology, 10*, 23.

Sharma, A., Kawarabayasi, Y., & Satyanarayana, T. (2012). Acidophilic bacteria and archaea: Acid stable biocatalysts and their potential applications. *Extremophiles, 16*, 1–19.

Shiratori, H., Sasaya, K., Ohiwa, H., Ikeno, H., Ayame, S., Kataoka, N., Miya, A., Beppu, T., & Ueda, K. (2009). Clostridium clariflavum sp. nov. and Clostridium caenicola sp. nov., moderatelythermophilic, cellulose-/cellobiose-digesting bacteria isolated from methanogenic sludge. *International Journal of Systematic and Evolutionary Microbiology, 59*, 1764–1770.

Shivaji, S., Bhadra, B., Rao, R. S., & Pradhan, S. (2008). Rhodotorula himalayensis sp. nov., a novel psychrophilic yeast isolated from Roopkund Lake of the Himalayan mountain ranges, India. *Extremophiles, 12*, 375–381.

Siddiqui, K. S., & Cavicchioli, R. (2006). Cold-adapted enzymes. *Annual Review of Biochemistry, 75*, 403–433.

Simonato, F., Campanaro, S., Lauro, F. M., Vezzi, A., D'Angelo, M., & Vitulo, N. (2006). Piezophilic adaptation: A genomic point of view. *Journal of Biotechnology, 126*, 11–25.

Singh, O. V., & Gabani, P. (2011). Extremophiles: Radiation resistance microbial reserves and therapeutic implications. *Journal of Applied Microbiology, 110*(4), 851–861. https://doi.org/10.1111/j.1365-2672.2011.04971.x

Singh, R., Dhawan, S., Singh, K., & Kaur, J. (2012). Cloning, expression and characterization of a metagenome-derived thermoactive/thermostable polygalacturonase. *Molecular Biology Reports, 39*(8), 8353–8361.

Sorokin, D. Y., & Muyzer, G. (2010a). Desulfurispira natronophila gen. nov. sp. nov.: An obligately anaerobic dissimilatory sulfur-reducing bacterium from soda lakes. *Extremophiles, 14*, 349–355.

Sorokin, D. Y., & Muyzer, G. (2010b). Haloalkaliphilic spore-forming sulfidogens from soda lake sediments and description of Desulfitispora alkaliphila gen. nov., sp. nov. *Extremophiles, 14*, 313–320.

Srivastava, R. A., Mathur, S. N., & Baruah, J. N. (1984). Partial purification and properties of thermostable intracellular amylases from a thermophilic Bacillus sp. AK-2. *Acta Microbiologica Polonica, 33*(1), 57–66.

Sthal, S. (1993). In M. N. Gupta (Ed.), *Thermostability of enzymes* (pp. 45–74). Springer.

Sukumaran, R. K., Singhania, R. R., Mathew, G. M., & Pandey, A. (2005). Cellulase production using biomass feed stock and its application in lignocellulose saccharification for bio-ethanol production. *Renewable Energy, 30*(12), 1479–1486.

Sunna, A., & Bergquist, P. L. (2003). A gene encoding a novel extremely thermostable 1,4-betaxylanase isolated directly from an environmental DNA sample. *Extremophiles, 7*, 63–70.

Sutrisno, A., Ueda, M., Abe, Y., Nakazawa, M., & Miyatake, K. (2004). A chitinase with high activity toward partially N-acetylated chitosan from a new, moderately thermophilic, chitin-degrading bacterium, Ralstonia sp. strain A-471. *Applied Microbiology and Biotechnology, 63*, 398–406.

Synowiecki, J. (2010). Some applications of thermophiles and their enzymes for protein processing. *African Journal of Biotechnology, 9*(42), 7020–7025.

Szilágyi, A., & Závodszky, P. (2000). Structural differences between mesophilic, moderately thermophilic and extremely thermophilic protein subunits: Results of a comprehensive survey. *Structure, 8*, 493–504.

Tamegai, H., et al. (2017). Extremophiles in the hadal trenches, the deepest oceanic environments. *FEMS Microbiology Ecology, 93*(6), fix057.

Tang, W., et al. (2019). A thermostable monoacylglycerol lipase from marine Geobacillus sp. 12AMOR1: Biochemical characterization and mutagenesis study. *International Journal of Molecular Sciences, 20*(3).

Takai K., Nakamura K., Toki T., Tsunogai U., Miyazaki M., Hirayama H., Nakagawa S., Nunoura T., Horikoshi K. (2008). Cell proliferation at 122°C and isotopically heavy CH4 production by a hyperthermophilic methanogen under high-pressure cultivation. *PNAS 105*, 10949–10954.

Taylor, I. N. R., Brown, C., Rycroft, M., King, G., Littlechild, J. A., Lloyd, M. C., et al. (2004). Application of thermophilic enzymes in commercial biotransformation processes. *Biochemical Society Transactions, 32*, 290–292.

Toplak, A., et al. (2013). Proteolysin, a novel highly thermostable and cosolvent-compatible protease from the thermophilic bacterium Coprothermobacter proteolyticus. *Applied and Environmental Microbiology, 79*(18), 5625–5632.

Tripathi, R., Singh, J., Bharti, R. K., & Thakur, I. S. (2014). Isolation, purification and characterization of lipase from microbacterium sp. and its application in biodiesel production. *Energy Procedia, 54*, 518–529.

Ueno, S., Kaieda, N., & Koyama, N. (2000). Characterization of a P-type Na+-ATPase of a facultatively anaerobic alkaliphile, Exiguobacterium aurantiacum. *The Journal of Biological Chemistry, 275*, 14537–14540.

Ugur, A., Sarac, N., Boran, R., Ayaz, B., Ceylan, O., & Okmen, G. (2014). New lipase for biodiesel production: Partial purification and characterization of LipSB 25-4. *International Scholarly Research Notices, 2014*, 289749.

Van der Maarel, M. J. E. C., van der Veen, B., Uitdehaag, J. C. M., Leemhuis, H., & Dijkhuizen, L. (2002). Properties and applications of starch-converting enzymes of the α-amylase family. *Journal of Biotechnology, 94*, 137–155.

Ventosa, A., Nieto, J. J., & Oren, A. (1998). Biology of moderately halophilic aerobic bacteria. *Microbiology and Molecular Biology Reviews, 62*(2), 504–544.

Vidyasagar, M., Prakash, S., Mahajan, V., Shouche, Y. S., & Sreeramulu, K. (2009). Purification and characterization of an extreme halothermophilic protease from a halophilic bacterium Chromohalobacter sp. TVSP101. *Brazilian Journal of Microbiology, 40*(1), 12–19.

Vieille, C., & Zeikus, J. G. (2001). Hyperthermophilic enzymes: Sources, uses, and molecular mechanisms for thermostability. *Microbiology and Molecular Biology Reviews, 65*, 1–43.

Viswanath, B., Chandra, M. S., Kumar, K. P., & Reddy, B. R. (2008). Production and purification of laccase from Stereum ostrea and its ability to decolorize textile dyes. *Dynamic Biochemistry, Process, and Biotechnology, 2*, 19.

Wang, L., Chi, Z. M., Wang, X. H., Liu, Z. Q., & Li, J. (2007). Diversity of lipase producing yeasts from marine environments and oil hydrolysis by their crude enzymes. *Annals of Microbiology, 57*, 495.

Wagner, I. D., & Wiegel, J. (2008). Diversity of thermophilic anaerobes. *Annals of the New York Academy of Sciences, 1125*, 1–43.

Wiegel, J., & Ljungdahl, L. G. (1996). The importance of thermophilic bacteria in biotechnology. *CRS-Reviews in Biotechnology, 3*, 39–107.

Wiegel, J., & Canganella, F. (2000). *Extreme thermophiles. Encyclopedia of Life Sciences*. Nature Publishing Group.

Yadav, A. N. (2021). Biodiversity and bioprospecting of extreme-philic microbiomes for agro-environmental sustainability. *Journal Of Applied Biology & Biotechnology, 9*(3), 1–6. https://doi.org/10.7324/JABB.2021.9301

Yadav, M., Shukla, A. K., Srivastva, N., Upadhyay, S. N., & Dubey, S. K. (2016). Utilization of microbial community potential for removal of chlorpyrifos: A review. *Critical Reviews in Biotechnology, 36*(4), 727–742.

Yao, J., Chen, Q., Zhong, G., Cao, W., Yu, A., & Liu, Y. (2014). Immobilization and characterization of tannase from a metagenomic library and its use for removal of tannins from green tea infusions. *Journal of Microbiology and Biotechnology, 24*(1), 80–86.

Zaccai, G. (2004). The effect of water on protein dynamics. Philosophical transactions of the Royal Society of London. *Series B: Biological Sciences, 359*, 1269–1275.

Zhang, G., Jiang, N., Liu, X., & Dong, X. (2008). Methanogenesis from methanol at low temperatures by a novel psychrophilic methanogen, "Methanolobus psychrophilus" sp. nov., prevalent in Zoige wetland of the Tibetan plateau. *Applied and Environmental Microbiology, 74*, 6114–6120.

Zhang, G. I., Hwang, C. Y., Kang, S. H., & Cho, B. C. (2009). Maribacter antarcticus sp. nov., a psychrophilic bacterium isolated from a culture of the Antarctic green alga Pyramimonas gelidicola. *International Journal of Systematic and Evolutionary Microbiology, 59*, 1455–1459.

Zhang, J., Shi, H., Xu, L., Zhu, X., & Li, X. (2015). Correction: Site-directed mutagenesis of a hyperthermophilic endoglucanase Cel12B from Thermotoga maritima based on rational design. *PLoS One, 10*(10), e0141937.

Zhu, S., et al. (2013). Biosynthesis of nucleoside analogues via thermostable nucleoside phosphorylase. *Applied Microbiology and Biotechnology, 97*(15), 6769–6778.

Zhu, D., Adebisi, W. A., Ahmad, F., Sethupathy, S., Danso, B., & Sun, J. (2020). Recent development of extremophilic bacteria and their application in biorefinery. *Frontiers in Bioengineering and Biotechnology, 8*, 483.

Ziemer, C. J. (2013). Newly cultured bacteria with broad diversity isolated from 8 week continuous culture enrichments of cow faeces on complex polysaccharides. *Applied and Environmental Microbiology, 80*(2), 574–585.

# Chapter 4
# Bioprospecting of Microalgae Derived Commercial Significant Compounds

**Contents**



## 4.1 Introduction

One of the main factors propelling the global nutraceuticals market is the growing concern over eating healthily. The market for nutraceuticals increased to $205.39 billion globally in 2016, and previously according to Ranga et al. (2017), it will rise at a compound annual growth rate (CAGR) of 6.5% from 2018 to 2023, reaching $297 billion. A recent report revealed that with an increased compound annual growth rate of 6.9%, the microalgal market (biomass or algal-based products) is predicted to produce up to USD 76.68 million by 2028 (Data Bridge Market Research, Citation, 2021). Although microalgae are less well-researched than seaweeds, they have advantages such as quick development, high photosynthetic efficiency, and the ability to be grown in production settings (Dolganyuk et al., 2020). *Spirulina* (*Arthrospira*) and *Chlorella*, in the form of dried biomass, were the most sought-after genera due to their numerous positive health effects in the global algae industry. *Arthrospira* is marketed exclusively based on its superior protein content around the world (Karkala et al., 2021). Protein and fat supplements that are

 
P. Saini, P. Mishra, *Bioprospecting of Microbial Resources for Agriculture, Environment and Bio-chemical Industry*, https://doi.org/10.1007/978-3-031-63844-2_4

normally processed can be replaced with whole microalgal biomass and its derivatives, which are considered to be best in the vegan diet. For the ensuing 10 years, powdered or pill forms of microalgal supplements are anticipated to rule the markets (Data Bridge Market Research, 2021). Currently, there are two primary uses for microalgae: the generation of biomass as a biologically active supplement and the cultivation of microalgae to isolate biologically active compounds from biomass (Dolganyuk et al., 2020). These bioactive compounds and nutrients, polysaccharides (specifically sulphatedpolysaccharides) including proteins, lipids, polyunsaturated fatty acids (PUFA), vitamins, pigments, phycocyanin, phycobiliproteins, carotenoids, enzymes, etc., are abundant in microalgae. Microalgae's biologically active compounds have anti-inflammatory, regenerative, anti-inflammatory, antihypertensive, antibacterial, antiviral, anticancer, and immunostimulant properties. Nevertheless, the majority of microalgal species have not yet been used on an industrial basis. Globally, there is an increasing need for microalgal metabolites in nutraceutical applications. Microalgae are very desirable because they are a natural supply of bioactive chemicals with a wide range of biological properties. These autotrophic organisms are a very diverse and wide group that can be unicellular or multicellular. As photosynthetic organisms, these groupings represent the base of the marine food chain and contribute significantly to the ocean's productivity. The most widely accessible microalgal products on the market are fatty acids, proteins, and pigments (Nagi et al., 2021). Substantial concentrations of PUFA are present in marine microalgae (Maneechote et al., 2021), particularly those from the n-3 series, such as docosahexaenoic and eicosapentaneoic acids (EPA and DHA, respectively). Just a small number of the majority of known microalgae species have been given the go-ahead to be used commercially as dietary supplements. These species are *Arthrospira, Chlorella, Crypthecodinium, Dunaliella* and *Odontella.* Microalgae are an important supply of carbon materials that can be utilized to make health supplements, medications, cosmetics, and biofuels. The microalgal business has been hindered by several issues related to the production of microalgae, including mass cultivation, strain improvement, stress adaptation, biomass disruption, nutrient variability, and water reprocessing. Microalgal biotechnology has the potential to bring about amazing discoveries and advancements which may result in integrated technologies for enormous production of microalgae-derived products.

## 4.2 Microalgae Characteristic Features

Microalgae are primarily unicellular, while some are multicellular. A class of autotrophic microorganisms known as microalgae inhabit freshwater, marine, and soil environments. Through photosynthesis, they create organic compounds. Prokaryotic cyanobacteria and eukaryotic algae make up microalgae. Studies on their usage as a source of physiologically valuable products are proliferating due to their great metabolic flexibility, ability to adapt to a variety of cultivation settings, and quick growth potential. Three basic strategies can be used to cultivate microalgae:

photoautotrophic, heterotrophic, and mixotrophic environments. The most popular culture technique is photoautotrophic. Since development does not always correspond with that of conventional algal strains, it is necessary to conduct experiments to establish the ideal growth parameters for each native strain found in a given habitat. The ideal circumstances for microalgae growth are contingent upon the light intensity, wavelength, and photoperiod that the cells are subjected to throughout the cultivation process. The photochemical phase of photosynthesis, during which light is absorbed by chlorophyll molecules, adenosine triphosphate (ATP) generation, and water photolysis, is directly correlated with light intensity (Dolganyuk et al., 2020). Cultured cells' growth rate is strongly impacted by the amount of light they receive since carbon is fixed. Photosynthetically active radiation (PAR) in the range from 400 to 700 nm from light radiation (corresponds to 50% of solar radiation) and intensity from 800 to 1000 $W/m^2$ is used by microalgae. Depending on the strain, the ideal growing temperature for the most often used microalgae, including *Chlorella, Chlamydomonas, Botryococcus, Scenedesmus, Neochloris, Haematococcus,* and *Nannochloropsis*, is between 15 and 35 °C. Although, it has been reported that the optimum temperature range for most of the microalgae culture is between 27 and 30 °C to maximize biomass yield. The ideal pH range for certain microalgae is between 6 and 8. Several instances have shown a continuous rise in the biomass of microalgae at very low pH levels. One such instance is *Chlorella protothecoides var. acidicola*, which was isolated from the microbial mat's surface and whose growth is optimal at pH 2.5 and 30 °C (Nancucheo & Johnson, 2012). In addition, microalgal growth medium must contain adequate amount of other nutrients such as Magnesium (Mg), Sulphur (S), Calcium (Ca), Sodium (Na), Chloride (Cl), Iron (Fe), Zinc (Zn), Copper (Cu), Molybdenum (Mo), Manganese (Mn), Boron (B) and Cobalt (Co), with an emphasis on Mg, S and Fe. Four methods are available for harvesting the microalga biomass: flotation, centrifugation, filtration, and biomass aggregation (flocculation and ultrasonic). In certain situations, multiple methods may be combined.

## 4.3 Microalgae Classification

Algae are found all over their ecosystem. They can grow in freshwater, brackish water, buildings, soils, and waters with different salinities. They can also coexist harmoniously with other living things. They are mainly separated into two categories: macroalgae, or larger algae, and microalgae, or smaller algae. Microalgae are tiny single-cell photosynthetic organisms that can be classified as either eukaryotic, like green algae (Chlorophyta), or prokaryotic, like cyanobacteria (Chloroxybacteria). They can be unicellular, multicellular, filamentous, or siphonaceous. One of the traits used to categorize the many kinds of algae is the presence of distinct pigments. For applied phycologists, algae taxonomy has remained a mystery. The description of microalgae species at a specific time and location is crucial to understanding how energy is transmitted from the base of the food chain to higher trophic levels, which makes microalgae taxonomy vital to biology and oceanology. Microalgae are very valuable both commercially and socially. Identification of microalgae species and growth monitoring are essential components in practical applications. Conventional manual techniques take a lot of time and require operators to have a high level of ability and experience. Data resources and computer processing power have increased dramatically in recent years. The development of microalgae classification, the conversion of microalgae to bioenergy, the use of microalgae for environmental protection, and the tracking of microalgae growth stages are all common applications for machine learning algorithms like support vector machines, decision trees, random forests, and neural networks (Ning et al., 2022). A recent study suggested a method for classifying microalgae by combining semi-supervised and active learning algorithms, which would produce a database of microalgae that could be analyzed by a FlowCAM particle analyzer (Drews et al., 2013).

## 4.4 Commercial Significant Compounds Derived from Microalgae

Microalgae have many potential uses in the biopharmaceutical, nutraceutical, and renewable energy sectors; they have garnered significant attention as treatment for water bioremediation recently microalgae can growth in wastewater on a global basis. Microalgae have the added benefit of inducing secondary metabolism, which opens the door for a variety of bioactive chemicals, because of their worldwide adaptation in any harsh environment and metabolic plasticity under stressed vs. non-stressed settings (Prarthana & Maruthi, 2019). Due to their constant exposure to a variety of stressors, including high or low temperatures, high salinities, osmotic pressure, photo-oxidation, and UV radiation, microalgae create and release useful compounds (Mishra & Prasad, 2021). Marine microalgae contain a wide range of biologically active substances, including phenolic compounds, halogenated ketones, cyclic polysulphides, amino acids, alkaloids, flavonoids, phlorotannins, carotenoids, and steroids from alkanes (Fig. 4.1).

Depending on the type of solvent utilized, different bioactive chemicals can be extracted. Different polar solvents extract phenolic compounds differently than non-polar solvents do so for flavonoids. Soxhlet extraction produces heat-stable components, whereas cold extraction produces a variety of lipids. Using organic solvents such as methanol, ethanol, and hexane and soxhlet extraction at 60 °C degrees Celsius, Prakash et al. assessed the antimicrobial activity of microalgae *Oscillatoria santa*, *Lyngbya birgeri*, *Oedogonium echinospermum*, *Spirogyra decimina*, *Spirogyra grantiana*, *Spirogyra crassa*, *Siprogyra bioformis*, and *Spirogyra condensate* against human pathogens such as *E. coli*, *Staphylococcus aureus*, *Salmonella typhimurium*, *Proteus vulgaris*, *Proteus mirabilis*, and *Streptococcus* infected.

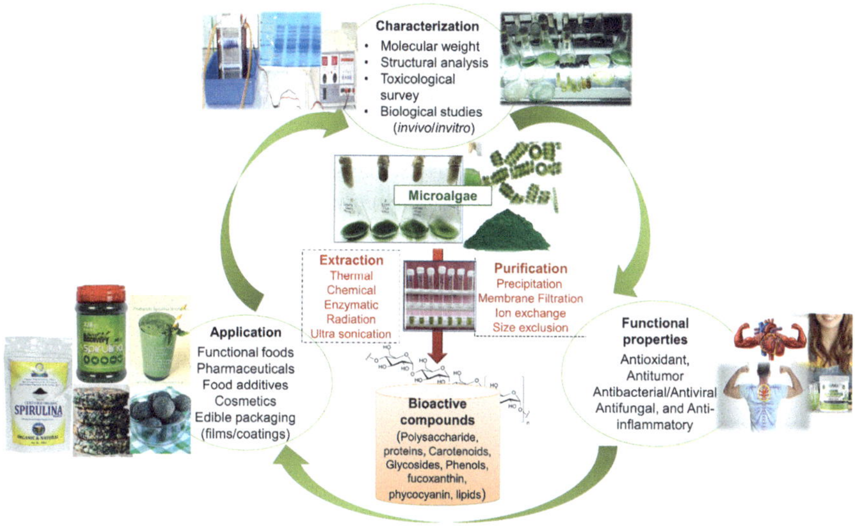


**Fig. 4.1** Schematic representation of microalgae-derived bioactive compounds with their potential application

Microalgae are inexpensive, sustainable, and renewable sources of nutritional components, bioactive pharmaceuticals, and biofuels. Even though these aquatic groups are identified by their poisons, microalgae offer a singular possibility for the discovery of novel bioactive chemicals. A lot of work is currently being done to develop new therapeutically helpful drugs. At the moment, a few items made from microalgae are on the market, and interest in microalgae biotechnology is growing. In some of their products, like PEPHA®-TIGHT CB or PEPHA®-CTIVE CB, labels like ©DSM use extracts from *Nannochloropsis oculata* and *Dunaliella salina* to promote cell proliferation and give skin-tightening effects, respectively (Martínez-Ruiz et al., 2022). Microalgae biomass is a prospective source of functional and nutritional supplements. Human consumption of microalgae is restricted to a few numbers of species, including *Dunaliella, Aphanizomenon, Nostoc, Chlorella* or *Chlorella vulgaris*, and *Arthrospira* (Spirulina) *platensis*. While they have established themselves as a source of beneficial components in aquaculture, feed, fertilizers, and cosmetics, other microalgae species like *Chlamydomonas* sp., *Chlorococcum* sp., *Scenedescmus* sp., *Tetraselmis chuii*, and *Nanochloropsis* sp. do not yet have GRAS (Generally recognised as safe) status. Moreover, the commercial production of astaxanthin and β-carotene are presently done from microalgae *Dunaliella salina* and *Hamatococcus pluvialis*, respectively while the commercially viable source for fucoxanthin is *Phaeodactylum tricornutum*.

Microalgae like *Odontella, Porphyridium cruentum, Arthrospira platensis, I. galbana* constitute a huge amount of lipids. Numerous studies have demonstrated the potential of microalgae biomass as a renewable energy source—specifically, biofuel. Over the past few decades, there has been a significant increase in the industrial cultivation of microalgae for the production of biofuels and bioproducts. *Chlamydomonas reinhardtii* and *Chlorella vulgaris* are two microalgae that are rich in carbohydrates and are thought to have the potential for techno-economic analysis (TEA) of bioethanol production. Microalgae produce polyunsaturated fatty acids, which are valuable commercial goods with strong therapeutic value for conditions like arthritis, asthma, and heart disease. These days, high-nutrient microalgae products like *Arthrospira, Chlorella*, and *Aphanizomenon flos-aquae* are accessible as extracts, pills, or capsules in their pure form as well as additions for use in a variety of food items like pastas, candies, gums, and drinks. Algae Systems, Kai BioEnergy, Pond Biofuels Incorporated, Cyanotech, and Algae Tec were a few of the major companies in the algae sectors.

## 4.5 Potential Benefits

With the advent of probiotic dietary supplements three to four decades ago, the health advantages of microalgae have come to be understood and appreciated more. It has been demonstrated that the microalga, or more precisely, the cyanobacteria *Athrospira* sp., increases the plasminogen activating factor in endothelial cells, which has a beneficial effect on the prevention of cardiovascular disease. Algae with antiviral, antioxidant, antifungal, and antibacterial properties are abundant in brown, red, and green pigment compounds (Kannan et al., 2014). Microalgae present a promising avenue for obtaining bioactive elements with potential applications in medicine. Table 4.1 highlights several significant components obtained from microalgae and their potential health benefits. Bioactive compounds with immunomodulatory activity, including amino acids, peptides, pigment-protein complex, and exopolysaccharides, are found in microalgae and cyanobacteria; of these, sulphated polysaccharides have been linked to blood coagulation, antilipidemic activity, and primarily immunomodulatory activity (Tabarzad et al., 2020). Additionally, this cyanobacterium has been shown to strengthen the immune system by preventing cancer and viral infections and expanding the flora of gastrointestinal microorganisms (Barrow & Shahidi, 2008). Because they contain useful metabolites that can be used to generate a wide range of varied uses in industries including agriculture, food, medicine, energy, and others, microalgae and cyanobacteria have attracted biotechnology interest. Stress variables can either boost or decrease the synthesis of various useful chemicals. As a result, scientific research aims to determine the ideal conditions for microalgae cultivation to maximize commercial production (Patel et al., 2019). The variety of algae offers a novel approach and can be recognized as an important source for the development of upcoming natural antiviral drugs. Marine macroalgae produce algal products (tannins, lectins, and polysaccharides) that have anti-HIV-1 properties. The marine blue-green algae *Arthrospira platensis* produces a sulfated polysaccharide called calcium spirulan (Ca-SP), which has been shown to suppress HIV-1 in MT-4 cells (Hayashi et al., 1996).

**Table 4.1** Potential bioactive compounds obtained from various microalgae

| Microalgae | Bioactive compounds | Potential benefits | References |
|---|---|---|---|
| *Griffithsia sp.* (Red algae) | Lectin, Griffithsin 12 | Antiviral activity against SARS-CoV, hepatitis C Infection, HSV-2 vaginal infection | O'Keefe et al. (2010) |
| *Porphyridium sp.* (Red algae) | Sulfated polysaccharide (carrageenan), xylose, glucose, galactose, and sulfate esters, hexuronic acids glucuronic acid and galacturonic acid | Herpes simplex virus HSV1 and HSV-2, varicela zoster virus (VZV), retrovirus murine sarcoma virus (MuSV-124), MuSV/MuLV (murine leukemia virus), hepatitis B virus (HBV), viral haemorrhagic septicaemia virus (VHSV), African swine fever virus (ASFV), vesicular stomatitis virus (VSV), vaccinia virus VACV and VACV-GFP and ectromelia virus (ECTV) | Gaikwad et al. (2020) |
| *Schizymenia pacifica* (Red algae) | Polysaccharidеn, sulfated galactan | Antiviral activity against herpesviruses (HSV types 1 and 2) | Matsuhiro et al. (2005) |
| *Gracilaria corticata* (Marine red algae) | Terpenoid, 4-acetoxy-2 hydroxy, 2, 6, 6-trimethyl cyclohexanone | Antiviral activity against herpes simplex virus types 1 and 2 | Govindasamy et al. (2012) |
| *Gracilaria lemaneiformis* (Red algae) | Sulfated polysaccharide | Anti-influenza virus | Chen et al. (2010) |
| *Dictiyota bartaysiana and Turbinaria decurrence* (Brown algae) | Sulphated polysaccharide, fucoidan | HIV (HIV-1 and HIV-2) | Sanniyasi et al. (2019) |
| *Isochrysis galbana* (Brown algae) | Carotenoids, Diatoxanthi, fucoxanthin | Antitumor, antibacterial, antioxidant properties, anti-inflammatory and hypochloesteromic | Mishra and Mishra (2018) |
| | lipolytic enzymes, 1-lyso-2-DHA-phospholipids, CCAP 927/1 | | |

(continued)

**Table 4.1** (continued)

| Microalgae | Bioactive compounds | Potential benefits | References |
|---|---|---|---|
| *Ecklonia stolonifera* (Brown algae) | Sterols and phlorotannins | Influenza Virus Neuraminidase Inhibitory Activity porcine epidemic diarrhea coronavirus infection and hemagglutination acetylcholine esterase inhibitory activity | Yoon et al. (2009), Wei et al. (2010), Ryu et al. (2011) |
| *Spirulina* (Blue green algae) | Protein, Griffithsin (GRFT) | Antiviral activity against SARS coronavirus (SARS-CoV) | https://www.sciencedaily.com/releases/2009/05/090520114659.htm |
| | Sulfated polysaccharide Alkaloids, Flavonoids, Glycosides, Phenols, Saponins, Cyanovirin-N (CV-N), Calcium spirulan (Ca-SP) –Antiviral 624olysaccharide | Antiviral activity against human immunodeficiency virus (HIV), hepatitis C virus (HCV), anti-hepatitis activity Immuno Modulating antioxidant potential | Nagulendran and Kavitha (2020) |
| *Caulerpa racemosa* (Green algae) | Alkaloids, phenolics, flavonoids andsteroids | Antiviral activity against against SARS-CoV | Srivastava et al. (2010), Lin et al. (2014) |
| *Padina gymnospora,* (Brown algae) | | | |
| *Palisada perforate* (Red algae) | | | |

Furthermore, certain species of microalgae may accumulate heavy metals and toxins, cause allergic reactions, or cause serious adverse effects. Therefore, to guarantee the safety, effectiveness, and quality of derived microalgal purified compounds and enriched extracts for approval and commercialization, a strong emphasis on good manufacturing practices in cultivation, harvesting, extracting, and purification as well as controls to limit toxin and impurities are needed.

## 4.6 Conclusion

Microgreen algae and cyanobacteria are typically identified in freshwater environments by their toxins. The range of nitrogen-rich alkaloids and peptides found in these toxins is frightening to humans, but it also holds great promise for the advancement of medicinal and agricultural uses. The solvent and extraction technique used has a significant impact on the exaction of bioactive substances. Moreover, a broad spectrum of biological activities, including ultraviolet (UV) absorption, antioxidant, anti-aging, anti-blemish, anti-inflammatory, and antibacterial qualities, are well acknowledged to be exhibited by microalgae metabolites. Furthermore, compared to higher plants, microalgae collect $CO_2$ more effectively, minimizing the greenhouse effect. Nonetheless, microalgae, particularly those belonging to marine species, are still mostly unexplored and offer a rare chance to investigate new secondary metabolites at a reasonable cost. Several researchers have effectively assessed the use of microalgae as a source of bioactive compounds. It is crucial to take into account the microalgae strain, growing conditions, biomass characterization, extraction, and functionalization techniques when utilizing the potential of microalgae-derived bioactive compounds to create and enhance a new generation of low-cost and eco-friendly commercial product.

## References

Barrow, C., & Shahidi, F. (2008). *Marine nutraceuticals and functional foods*. CRC Press; United State.

Chen, M. -Z., Xie, H. -G., Yang, L. -W., Liao, Z. -H., & Yu, J. (2010). In vitro anti-influenza virus activities of sulfated polysaccharide fractions from Gracilaria lemaneiformis[J]. *Virol Sin, 25*(5), 341–351.

Data Bridge Market Research. (2021). *Global microalgae market—Industry trends and forecast to 2028*. Accessed on: 23 Apr 2021, from https://www.databridgemarketresearch.com/reports/global-microalgae-market

Dolganyuk, V., Belova, D., Babich, O., Prosekov, A., Ivanova, S., Katserov, D., Patyukov, N., & Sukhikh, S. (2020). Microalgae: A promising source of valuable bioproducts. *Biomolecules, 10*(8), 1153.

Drews, P., Colares, R. G., Machado, P., et al. (2013). Microalgae classification using semi-supervised and active learning based on Gaussian mixture models. *Journal of the Brazilian Computer Society, 19*, 411–422.

Gaikwad, M., Pawar, Y., Nagle, V., & Dasgupta, S. (2020). *Marine red alga Porphyridium sp. as a source of sulfated polysaccharides (SPs) for combating against COVID-19.* Reliance Industries Ltd., RCP, Ghansoli, Thane-400701 Preprints (www.preprints.org).

Govindasamy, C., Arulpriya, M., & Ruban, P. (2012). Nuclear magnetic resonance analysis for antimicrobial compounds from their seaweed Gracilaria corticata. *Asian Pac J Trop Biomed, 2*, S329–S333.

Hayashi, K., Hayashi, T., & Kojima, I. (1996). A natural sulfated polysaccharide, calcium spirulan, isolated from *Spirulina platensis*: In vitro and ex vivo evaluation of anti-herpes simplex virus and anti-human immunodeficiency virus activities. *AIDS Research and Human Retroviruses, 12*, 1463–1471.

Judy George, Senior Staff Writer, MedPage Today April 28, 2020 (n.d.). *Young COVID-19 Stroke Victims; Alzheimer's Seaweed Drug; Parkinson's Med OK'd—News and commentary from the world of neurology and neuroscience* https://www.medpagetoday.com/neurology/generalneurology/86190

Kannan, M., Pushparaj, A., Dheeba, B., Nageshwari, K., & Kannan, K. (2014). Phytochemical screening and antioxidant activity of marine algae *Gracilaria corticata* and *Spirulina platensis*. *Journal of Chemical and Pharmaceutical Research, 6*(11), 312–318.

Karkala, S., D'Souza, L., & Nivas, S. (2021). Bioprospecting microalgae harnessed from the coastal belt of Mangalore, India as prospective nutraceutical and biofuel candidates. *Applied Phycology, 2*(1), 60–73.

Lin, L. T., Wen, C. H., & Lin, C. C. (2014). Antiviral natural products and herbal medicines. *Journal of Traditional and Complementary Medicine, 4*(1), 24–35.

Matsuhiro, B., Conte, A. F., Damonte, E. B., et al. (2005). Structural analysis and antiviral activity of a sulfated galactan from the red seaweed Schizymenia binderi (Gigartinales, Rhodophyta). *Carbohydrate Research, 340*(15), 2392–2402.

Maneechote, W., Cheirsilp, B., Srinuanpan, S., & Pathom-aree, W. (2021). Optimizing physicochemical factors for two-stage cultivation of newly isolated oleaginous microalgae from local lake as promising sources of pigments, PUFAs and biodiesel feedstocks. *Bioresource Technology Reports, 15*, 100738.

Martínez-Ruiz, M., Martínez-González, C. A., Kim, D. H., Santiesteban-Romero, B., Reyes-Pardo, H., Villaseñor-Zepeda, K. R., Meléndez-Sánchez, E. R., Ramírez-Gamboa, D., Díaz-Zamorano, A. L., Sosa-Hernández, J. E., Coronado-Apodaca, K. G., Gámez-Méndez, A. M., Iqbal, H. M. N., & Parra-Saldivar, R. (2022). Microalgae bioactive compounds to topical applications products-a review. *Molecules, 27*(11), 3512.

Mishra & Mishra (2018). Exploring the biologically active metabolites of Isochrysis Galbana in Pharmaceutical interest: an overview. *Int J Pharma Sci Res, 9*(6), 2162–2174.

Mishra, P., & Prasad, S. M. (2021). Low dose UV-B radiation induced mild oxidative stress impact on physiological and nutritional competence of Spirulina (*Arthrospira*) species. *Plant Stress, 2*, 100039.

Nagi, G. K., Minhas, A. K., Gaur, S., Jain, P., & Mandal, S. (2021). Integration of algal biofuels with bioremediation coupled industrial commodities towards cost-effectiveness [review]. *Frontiers in Energy Research, 9*, 489.

Nagulendran, K., & Kavitha, N. (2020). Antiviral and immuno modulating role of super food spirulina and Covid-19. *International Journal of Trend in Scientific Research and Development, 4*(4), 621–626.

Nancucheo, I., & Johnson, D. B. (2012). Acidophilic algae isolated from mine-impacted environments and their roles in sustaining heterotrophic acidophiles. *Frontiers in Microbiology, 3*, 325.

Ning, H., Li, R., & Zhou, T. (2022). Machine learning for microalgae detection and utilization. *Frontiers in Marine Science, 9*, 947394.

O'Keefe, B. R., Giomarelli, B., Barnard, D. L., Shenoy, S. R., Chan, P. K. S., McMahon, J. B., Palmer, K. E., Barnett, B. W., Meyerholz, D. K., Wohlford-Lenane, C. L., et al. (2010). Broad-spectrum in vitro activity and in vivo efficacy of the antiviral protein griffithsin against emerging viruses of the family Coronaviridae. *Journal of Virology, 84*, 2511–2521.

Patel, A., Matsakas, L., Rova, U., & Christakopoulos, P. (2019). A perspective on biotechnological applications of thermophilic microalgae and cyanobacteria. *Bioresource Technology, 278*, 424–434.

Prarthana, J., & Maruthi, K. R. (2019). Fresh water algae as a potential source of bioactive compounds for aquaculture and significance of solvent system in extraction of antimicrobials. *Journal of Asian Scientific Research, 12*, 18–28.

Ranga, R. A., Vijaya, R. D., & Ravishankar, G. (2017). Secondary metabolites from algae for nutraceutical applications. *Novel Techniques in Nutrition & Food Science, 1*(1), NTNF.000503.

Ryu, Y. B., Jeong, H. J., Yoon, S. Y., Park, J. Y., Kim, Y. M., Park, S. J., Rho, M. C., Kim, S. J., & Lee, W. S. (2011). Influenza virus neuraminidase inhibitory activity of phlorotannins from the edible brown alga Ecklonia cava. *J. Agric. Food Chem. 59*, 6467–6473.

Sanniyasi, E., Venkatasubramanian, G., & GopalR, K. (2019). In vitro anti-HIV-1 activity of the bioactive compound extracted and purified from two different marine macroalgae (seaweeds) (Dictyota bartayesiana J.V.Lamouroux and Turbinaria decurrens Bory). *Scientific Reports, 9*, 12185.

Srivastava, N., Saurav, K., Mohanasrinivasan, V., Kannabiran, K., & Singh, M. (2010). Antibacterial potential of macroalgae collected from the Madappam coast, India. *British Journal of Pharmacology and Toxicology, 1*(2), 72–76.

Tabarzad, M., Atabaki, V., & Hosseinabadi, T. (2020). Anti-inflammatory activity of bioactive compounds from microalgae and cyanobacteria by focusing on the mechanisms of action. *Molecular Biology Reports, 47*, 6193–6205.

Wei, D., Wang, L., Liu, C., & Wang, B. (2010). Sitosterol solubility in selected organic solvents. *Journal of Chemical & Engineering Data, 55*, 2917–2919.

Yoon, N. Y., Lee, S.-H., & Kim, S.-K. (2009). Phlorotannins from Ishige okamurae and their acetyl- and butyrylcholinesterase inhibitory effects. *Journal of Functional Foods, 1*, 331–335.

# Chapter 5
# Biofilm Linked Microbial Prospecting of Bioremediation

**Contents**



## 5.1 Introduction

Rapid urbanization, industrial growth, and technological progress have significantly contributed to a notable rise in environmental pollution worldwide, caused by various harmful synthetic substances (Mohapatra et al., 2017). In recent times, the contamination of water, air, soil, and sediment with toxic heavy metals and industrial dyes (such as azo dyes and synthetic dyes) has become a cause for concern in our ecological system (Sinha et al., 2018; Shukla et al., 2017). Heavy metals, known for



P. Saini, P. Mishra, *Bioprospecting of Microbial Resources for Agriculture, Environment and Bio-chemical Industry*, https://doi.org/10.1007/978-3-031-63844-2_5

their extreme toxicity even at minimal concentrations and their resistance to biodegradation compared to organic pollutants, pose a persistent threat to both life and the environment. The primary origins of heavy metals and dyes are linked to diverse industrial activities, including mining, metallurgy, electroplating, electrolysis, electro-osmosis, electrical appliances, energy generation, surface finishing, distilleries, photography, tanneries, iron and steel production, pesticide and fertilizer manufacturing, atomic energy installations, as well as the manufacture of paints, textiles, rubber, plastics, paper, leather, cosmetics, food and pharmaceuticals (Shukla et al., 2014; Mohapatra et al., 2017). The presence of metal contaminants has been associated with adverse health effects, such as birth defects, skin issues, cancer, physical and mental developmental delays, learning disabilities, and damage to the kidneys and liver among other health issues.

The ongoing manufacturing and release of these pollutants into the environment are causing their accumulation at various levels of the food chain, as highlighted by Shukla et al. (2014). Consequently, the need for effective remediation methods has become imperative to mitigate these issues. Biological remediation technologies are favored over various physico-chemical approaches for removing metals due to their environmental safety, practical feasibility, and the absence of harmful by-products. Microorganisms have the ability to absorb and sequester metals within their cells, facilitating their removal from contaminated environments. Therefore, microorganisms play a crucial role in addressing widespread contamination, offering diverse applications (Delangiz et al., 2020). In comparison to other microbial bioremediation processes, techniques involving bacterial biofilm-mediated bioremediation show greater promise. This is attributed to their higher microbial biomass densities, immobilization capacity, and enhanced protective and adhesive properties (Shukla et al., 2017; Gupta & Diwan, 2017; Edwards & Kjellerup, 2013; Das et al., 2017). The effectiveness of biofilm organisms in bioremediation is primarily associated with efficient gene transfer within the biofilm species and bacterial chemotaxis, which enhances the bioavailability of contaminants (Singh et al., 2006).

A biofilm refers to a cohesive, adhesive layer with a slimy texture, characterized by a negative charge and composed of organized microbial communities consisting of either a single species or multiple species (mixed species). These communities adhere to surfaces, whether living (biotic) or non-living (abiotic), by becoming embedded in a self-produced matrix of extracellular polymeric substances (EPSs). These EPSs are complex mixtures of various high molecular weight biopolymers, including polysaccharides, proteins, nucleic acids, lipids, uronic acids, and humic substances (Gupta & Diwan, 2017; Di Donato et al., 2016). The secretion of EPSs by the microbes within the biofilm serves multiple purposes. Not only does it facilitate cell growth, but it also creates water channels that enable the transport of nutrients and oxygen (Shukla et al., 2014; Singh et al., 2006). The net negative charge of EPSs allows these biopolymers to efficiently capture positively-charged heavy metal ions and dye substances (Shukla et al., 2014, 2017). Additionally, EPSs contain microbial surface-active metabolites known as biosurfactants, which act as metal complexing agents. These biosurfactants play a crucial role in accessing water-insoluble hydrophobic or other resistant substrates in an environmentally compatible and biodegradable manner during the remediation process (Shukla et al., 2014, 2017; Mitra & Mukhopadhyay, 2015; Mitra et al., 2014a, b; Singh et al., 2006).

Microbes within a biofilm coexist and derive mutual benefits by creating diverse ecological niches. This communal living provides protection to groups of bacteria against challenging environmental conditions, including fluctuations in nutrition, exposure to toxic chemicals, variations in salt concentrations, pH levels, and temperature extremes (Vaishnav et al., 2016). Notably, biofilms exhibit an exceptionally high tolerance to hazardous inorganic elements, even when exposed to lethal doses. They thrive in humid environments with ample nutrient availability and a suitable surface for adherence. Typically, bacteria adhere to surfaces in a structured manner within a biofilm (Medfu Tarekegn et al., 2020; Sharma et al., 2019). This chapter delves into the advantageous applications of biofilm-producing bacteria in the bioremediation of environmental pollutants, such as heavy metals and dyes. It explores various aspects of bacterial biofilm development, the factors influencing this development, the interactions between bacterial biofilms and toxic metals and dyes, the mechanisms behind developing resistance to toxic pollutants, and sustainable strategies for biofilm-based bioremediation.

## 5.2 Biofilms

Microorganisms have traditionally been identified as planktonic, meaning they exist as freely suspended cells. Their classification has been based on morphological and physiological characteristics, as well as growth patterns in nutrient-rich environments. However, in wet and nutrient-abundant surroundings, microorganisms are commonly found closely associated with surfaces, forming multicellular clusters. This adherence results in the creation of sticky and viscous layers known as microbial slimes, which eventually develop into biofilms. Biofilms are aggregations of microbial cells firmly attached to surfaces and are prevalent in moist environments with sufficient nutrient availability and suitable surfaces for attachment. These biofilms can be composed of a single bacterial species or a variety of bacteria, fungi, algae, and protozoa. Interestingly, around 97% of the biofilm matrix consists of water, bound to microbial cell capsules, or solvent, whose physical properties, such as viscosity, are influenced by dissolved solutes (Watnick & Kolter, 1999; Sutherland, 2001). Biofilms can comprise a single microbial species or multiple microbial species and can form on a range of biotic and abiotic surfaces. Once an initial biofilm is formed, cell to cell communication (i.e. quorum sensing) through extracellular signaling molecules controls further modification and development of the biofilm (Kjelleberg & Molin, 2002). The diffusion processes taking place within the biofilm matrix are contingent upon both the water-binding capacity and mobility of the biofilm. Aside from water and microbial cells, the biofilm matrix constitutes an intricate blend of secreted polymers, assimilated nutrients, metabolites, cell lysis byproducts, and various particulate materials sourced from the adjacent environment (Flemming & Wingender, 2010). In this context, a diverse array of macromolecules—including proteins, polysaccharides, DNA, and RNA—coexists within the biofilm environment, in addition to conventional cellular components such as peptidoglycan, lipids, and phospholipids (Sutherland, 2001).

The anionic characteristics of biofilms arise from the presence of uronic acids such as D-glucuronic, D-galacturonic, and mannuronic acids, or ketal-linked pyruvates (Flemming, 1995). Mass transport within biofilms is shaped by the structure of the biofilm, a structure influenced by the localized availability of substrates. Solute transport in biofilms encompasses convective transport within pores and water channels, as well as diffusion in denser aggregates (Horn & Morgenroth, 2006). Consequently, the biofilm matrix exhibits a pronounced degree of microheterogeneity due to the coexistence of multiple microenvironments within it. Originally, biofilms were perceived as uniform collections of cells trapped in slime. However, recent research into bacterial biofilms has unveiled their physiological and genetic diversity. The biofilm matrix is primarily composed of water, microorganisms, secreted biopolymers, absorbed nutrients, metabolites, cell lysis products, and particulate materials from the nearby environment. The intricate structure of the biofilm matrix is largely contributed to by major cellular components such as peptidoglycan, phospholipids, lipids, proteins, polysaccharides, nucleic acids (DNA and RNA), uronic acids (D-galacturonic, D-glucuronic, and mannuronic acids), and ketal-linked pyruvates (Singh et al., 2006).

Bacterial biofilms offer several advantages compared to free-floating planktonic bacterial species. These benefits include protection against the adverse effects of environmental changes (e.g., pH, salinity, temperature, water flow, or nutrient levels), the ability to exchange nutrients and genetic material, survival in different metabolic states, and increased tolerance to various toxic compounds such as chemicals, organic pollutants, heavy metals, and antibiotics (Shukla et al., 2017; Edwards & Kjellerup, 2013; Das et al., 2017). Biofilms naturally thrive in diverse contaminated environments, reducing environmental stress by developing resistance and meeting each other's metabolic needs. Consequently, biofilms possess characteristics that enable them to withstand and adapt to harsh polluted conditions, making them valuable for activities like bioremediation, biotransformation, bioimmobilization, and the degradation of toxic xenobiotic compounds.

## 5.3 Composition of Biofilm

A biofilm is a cohesive and adhesive layer characterized by a negative charge, stickiness, and a slimy texture, formed by the collective associations of single or diverse microbial species.Biofilmsare made up of extracellular polymeric substances (EPS) made of proteins including polysaccharides (1–2%), enzymes, DNA (<1%), RNA (<1%), and in addition to these components, water is the major part of biofilm which is responsible for the flow of nutrients inside biofilm matrix (Kumar et al., 2009). The architecture of biofilm consists of two main components i.e. water channel for nutrients transport and a region of densely packed cells having no prominent pores in it. Bacteria are well known for their ability to produce a wide variety of polysaccharides classified according with their location relative to the cell. Those that are excreted outside the cell wall are called extracellular polysaccharides or EPSs, which can be tightly linked to the cell surface forming a capsular polysaccharide (CPS) or secreted as exopolysaccharides (EPS). Exopolysaccharides have a high molecular weight and biodegradable polymers formed by monosaccharide residues of sugar and sugar derivatives, and are produced by a wide range of bacteria. Exopolysaccharides (EPS) are large and biodegradable polymers composed of sugar and sugar derivatives. EPS play a crucial role in the formation of the biofilm matrix, as well as in mechanisms related to cell aggregation and adhesion, both on abiotic and biotic surfaces such as the intestinal mucosa (Whitfield & Valvano, 1993). In microbial life, EPS serves various ecological functions, enhancing bacterial colonization in technological or gastrointestinal environments. It facilitates stable cell recognition and cooperation, acting as a protective barrier against harmful substances like antibiotics and toxic compounds. This protective role ultimately contributes to increased bacterial survival (Flemming et al., 2016). Apart from forming hydrophilic molecules like DNA, polysaccharides, and proteins, extracellular polymeric substances (EPS) also generate hydrophobic compounds such as lipids and polysaccharides containing alkyl groups (Mitra et al., 2014a, b). These compounds serve as metal complexing agents, facilitating the acquisition of water-insoluble or otherwise challenging substrates in an environmentally friendly manner during removal processes. Biofilms contain a variety of enzymes that play a crucial role in dispersing the biofilm matrix and releasing microbial cells during relocation (Mangwani et al., 2017). Extracellular DNA, a significant component of the biofilm matrix, consists of high molecular weight fragments, some reaching up to 30 kb. This DNA can chelate cations, triggering resistance mechanisms in various bacterial strains. It is associated with heightened antibiotic resistance, the exchange of genetic information, and the ability of individual cells to self-organize and maintain mechanical stability within biofilms. These characteristics highlight the multifaceted role of extracellular DNA in biofilm dynamics and bacterial adaptability. The microbial cells within biofilms are arranged with significant different physiology and physical properties. Bacterial biofilms are normally beyond the access of antibiotics and human immune system. Microorganisms that produce biofilm have enhanced potential to bear and neutralize antimicrobial agents and result in prolonged treatment. Biofilm forming bacteria switch on some genes that activate the expression of stress genes which in turn switch to resistant phenotypes due to certain changes e.g. cell density, nutritional or temperature, its pH and osmolarity (Fux 2005).

## 5.4 Development of Biofilm

Biofilms can naturally develop in living organisms (in vivo) or be artificially created on surfaces within a laboratory setting (in vitro). These intricate and multicellular structures exist on both natural and artificial solid surfaces, encompassed by a network of water channels (Shukla et al., 2017). Biofilm is an association of micro-organisms in which microbial cells adhere to each other on a living or non-living surface within a self-produced matrix of extracellular polymeric substance. Biofilm formation is a multi-step process starting with attachment to a surface then formation of micro-colony that leads to the formation of three-dimensional structure and finally ending with maturation followed by detachment. During biofilm formation many species of bacteria are able to communicate with one another through specific mechanism called quorum sensing. It is a system of stimulus to co-ordinate different gene expression (Jamal et al., 2018). Biofilms are multicellular communities held together by a self-produced extracellular matrix. The mechanisms that different bacteria employ to form biofilms vary, frequently depending on environmental conditions and specific strain attributes. The study emphasized on four well-studied model systems to give an overview of how several organisms form biofilms: *Escherichia coli*, *Pseudomonas aeruginosa*, *Bacillus subtilis* and *Staphylococcus aureus*. Using these bacteria as examples, the key features of biofilms as well as mechanisms by which extracellular signals trigger biofilm formation were discussed (López & Kolter 2010).

The biofilm formation activity by various lactic acid bacteria isolated from honey bee guts or fresh pollen samples in the presence of different sugars added to the cultivation media was studied. The strongest biofilm formation ($OD_{570} \geq 1$) was noted in *A. kunkeei* V18 in the presence of fructose (Bielik et al., 2021). The influence of flow on the formation and growth of biofilms produced by gram-positive *Lactobacillus plantarum* strains WCFS1 and CIP104448 was investigated by Rashtchi et al. (2022) throughan in-house designed device based on a 48-well plate and quantified total biofilm formation under static and flow conditions using crystal violet (CV) staining. Plate counts showed that flow caused an increase in the number of viable biofilm cells for both strains. Biofilm formation as quantified by crystal violet staining and colony forming units is largely affected by the medium composition, growth temperature, maturation time and by strain specific features. All strains showed high biofilm formation in Brain Heart Infusion medium supplemented with manganese and glucose. The strain to strain variation may be explained by differences in levels of eDNA, likely as result of differences in lysis behaviour. The study showed the impact of a range of environmental factors and enzyme treatments on biofilm formation capacity for selected *L. plantarum* isolates (Fernández Ramírez et al., 2015). Presently, the consensus model in biofilm research advocates that microbial community development on surfaces is a stepwise process involving adhesion, growth, motility and EPS production (Molin & Tolker-Nielsen 2003). Once an initial biofilm is formed, cell to cell communication (i.e. quorum sensing) through extracellular signaling molecules controls further modification and development of the biofilm. Recently an *In vitro* model of microbial biofilm formation on polyvinyl chloride biomaterial (PVC) has been reported which was found to be a dynamic process with maximum thickness being attained after 12–18 h (Kjelleberg & Molin 2002).

## 5.5 Biofilm Formation Mechanisms

The development of bacterial biofilms involves a series of stages, beginning with reversible adhesion to surfaces facilitated by intermolecular interactions and hydrophobicity. This process then advances to the generation of extracellular polymeric substances (EPS), enabling the cells to adhere to a surface persistently (Caruso et al., 2018). Recent investigations have unveiled the gradual progression of bacterial biofilm formation, encompassing steps such as adhesion, growth, motility, and EPS production, as well as the modification and development of biofilms through mechanisms like quorum sensing and extracellular signaling (Shukla et al., 2014, 2017). Detailed stages of biofilm formation and development are described below (Fig. 5.1).

The initial stages of microbial community attachment to both living (biotic) and non-living (abiotic) surfaces involve adhesion and adsorption, influenced by the physico-chemical characteristics of microbial cells. The adhesion in biofilm

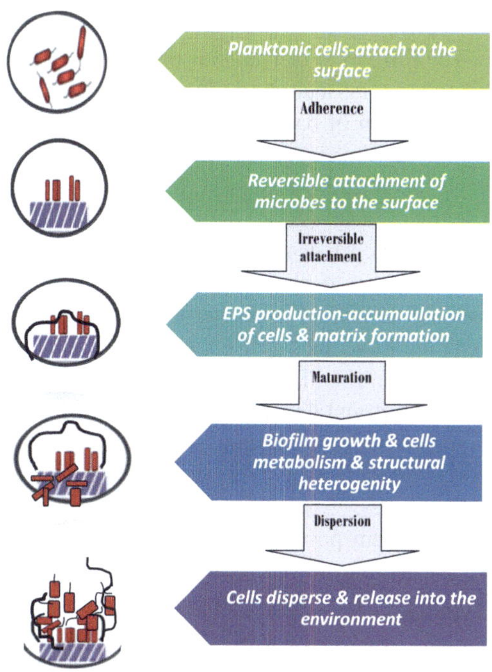


**Fig. 5.1** Formation and development of Biofilm

formation is primarily associated with various cell adhesion forces, such as the "Van der Waals force of interaction and hydrophobic interaction," attributed to hydrogen bonding (Shukla et al., 2017). The secretion of organic and inorganic substances from bacterial surfaces plays a crucial role in mitigating the electrostatic force of repulsion resulting from the identical charges on the base surface and bacterial cell, facilitating the irreversible adhesion of the biofilm (Mohapatra et al., 2018).

The irreversible attachment of the biofilm to the surface is achieved through bacterial surface secretions, which include both organic and inorganic molecules. This secretion helps diminish the electrostatic force of repulsion arising from the similar charges between the bacterial cell and the base surface (Shukla et al., 2017). Bacterial colonization within the biofilm involves the rapid adsorption of various macromolecules onto the surface to fulfill their nutritional requirements. As bacterial cells grow and multiply within the biofilm, accompanied by the production of extracellular polymeric substances (EPS), mechanical stability is achieved. A mature biofilm exhibits functional homogeneity and structural integrity akin to a quasi-tissue, with specific activities such as respiration, nutrient uptake, physiological cooperation, and self-protection (Das et al., 2017). Bacterial colonization within biofilms involves the utilization of a diverse range of macromolecules, rapidly adsorbing them onto the surface to fulfill nutritional requirements. As bacterial cells undergo growth and multiplication within the biofilm, they concurrently produce extracellular polymeric substances (EPS), contributing to the mechanical stability of the biofilm. Mature biofilms exhibit both structural integrity and functional homogeneity, resembling quasi-tissues in terms of activities such as respiration, nutrient uptake, self-protection, and physiological cooperation (Mohapatra et al., 2019).

The formation of bacterial biofilms is intricately regulated by various environmental factors, including pH variations, nutrient and oxygen availability, and the concentration of bacterial metabolites. These factors contribute to the distinctiveness of biofilms, causing variations between biofilms formed in different environments, such as those formed in streams versus those formed on biological tissues. Allegrucci et al. (2006) supported this notion, demonstrating that environmental factors influence de novo protein production within Streptococcus pneumoniae biofilms, leading to changes in cell number and biomass. Quorum sensing, which involves the ability of bacteria to communicate and coordinate behavior through signaling molecules, plays a crucial role in regulating changes within mature biofilms. It controls the formation of channels and pillar-like structures that enhance efficient nutrient delivery to the cells. This architectural arrangement facilitates the proper dispersal of nutrients, particularly when there is heightened competition for food at high cell density within biofilms (Stanley & Lazazzera, 2004; Kjelleberg & Molin, 2002).

Both Gram-positive and Gram-negative bacteria commonly utilize quorum sensing for cellular communication to govern a wide range of physiological activities, including symbiosis, competence, motility, sporulation, virulence, conjugation, antibiotic production, and biofilm formation (Miller & Bassler, 2001). Recent research has clarified that the formation and development of biofilms are strongly influenced by quorum sensing mechanisms, involving cell-to-cell communication through autoinducers, occurring both within and between bacterial species (Di Donato et al., 2016). In Staphylococcus aureus, the quorum-sensing system's cyclic-peptide-dependent accessory gene regulator (agr) plays a role in repressing several surface adhesins responsible for interacting with the host matrix. This repression facilitates the attachment of Staphylococcus aureus to various surfaces (Parsek & Greenberg, 2005; Kjelleberg & Molin, 2002). In the case of Vibrio cholerae, an enteric pathogen, quorum sensing is involved in regulating the production of secreted exopolysaccharides (EPS) encoded by the vps operon. These exopolysaccharides mediate intercellular aggregation and adherence to surfaces (Parsek & Greenberg, 2005). Studies conducted by Stanley et al. (2003), utilizing confocal scanning laser microscopy, observed that glucose inhibits biofilm formation in Bacillus subtilis through the catabolite control protein (CcpA). Despite these advancements, research on differential gene expression in bacterial biofilms is still in its early stages. Scientists are actively exploring a 'universal' biofilm gene expression pattern, the regulation of which could potentially guide the formation or destruction of biofilms.

## 5.6 Biofilm Linked Remediation

Bioremediation is a rising in-situ technology that employs microorganisms to clean up environmental pollutants. The biological processes utilized in treating toxic effluents demonstrate superior efficiency and cost-effectiveness compared to chemical and physical methods (Paul et al., 2005). Recently, the potential of biofilm communities for bioremediation processes has come to light. Biofilm-mediated bioremediation offers a proficient and safer alternative to traditional bioremediation using planktonic microorganisms. This is because cells within a biofilm have a higher likelihood of adaptation and survival, especially during periods of stress, as they are shielded within the biofilm matrix (Decho, 2000).

Compared to other emerging bioremediation methods, the use of bacterial biofilm for wastewater treatment has demonstrated numerous advantages. In the treatment of industrial effluents, biofilm is currently under investigation within various bioremediation systems, including biofilters, biostimulation, phytoremediation, aerobic and anaerobic granular sludge reactors, as well as rotating disk contractors (Sharma et al., 2020). Microbiologists across diverse fields, including agriculture, industry, environment and health are increasingly recognizing the significance of microbial biofilms. These biofilms play a crucial role in facilitating the proliferation and colonization of microbes on surfaces, simultaneously offering protection to cells from external environmental factors. Biofilms exhibit a range of physicochemical and biological processes that enable them to bind and transform toxic pollutants, including metals and dyes, utilizing extracellular polymeric substances (EPSs) and biosurfactants. The polyionic characteristics of biofilm EPS play a crucial role in binding metal ions, leading to the formation of organometal complexes through the electrostatic force of attraction (Shukla et al., 2017).

Biofilm-mediated bioremediation is further facilitated by mechanisms such as enhanced intercellular gene transfer, the use of mixed population biofilms, improvement of pollutant bioavailability to biofilm-residing bacterial cells through chemotaxis, and optimization of physicochemical parameters. The diffusion and transportation of various metal ions, solutes, and dye molecules through the biofilm matrix's pores and water channels are significantly influenced by the anionic property, water-binding capacity, cellular densities, and mobility of the biofilm. The microheterogeneity within the biofilm matrix, resulting from the coexistence of diverse microenvironments, enhances its potential for applications in bioremediation processes (Edwards & Kjellerup, 2013; Singh et al., 2006). Remediating through biofilm processes offers diverse approaches. Natural attenuation relies on inherent processes, not human interventions, where specific strains aid bioremediation by transforming contaminants into less harmful forms. For instance, microbial biofilm communities in soil can naturally convert pollutants. This method operates on the premise that certain pollutants can degrade, convert, immobilize, and detoxify without direct human intervention, requiring a resident profile of microorganisms capable of decomposing contaminants in biofilm mode, which is time-consuming. Biostimulation involves providing nutrition, growth substrates, and stimuli to increase bacterial activity for efficient pollutant biodegradation. On the other hand, bioenhancement or bioaugmentation involves introducing specific competent bacteria or microbial consortia to enhance degradation. Bioaugmentation is the favored approach for sites recently contaminated, where the natural microbial population may be insufficient or has diminished capacity for effective degradation. This setting also encourages the formation of microbial biofilms, enhancing the overall efficiency of degradation processes. This approach can be refined through genetic engineering, strategies to boost nutrients, microbial persistence, air venting, and additional biostimulation techniques (Liu et al., 2011; Tyagi et al., 2011; Gentry et al., 2004).

Bacterial biofilms can effectively be utilized in ex situ conditions through various engineered bioreactor systems. Different types of bioreactors, including fixed bed reactors, packed bed reactors, and fluidized bed reactors, leverage bacterial biofilm biomass for the treatment and biosorption of contaminated water containing heavy metals, hydrocarbons and dyes (Mitra & Mukhopadhyay, 2015). Recently discovered, microbes enveloped in biofilms exhibit significant bioremediation potential, forming mutually beneficial associations of multiple microorganisms in breaking down pollutants at industrial sites. Biofilm-associated microorganisms find application in various bioreactors and biofilters for large-scale pollution reduction (Rastegari et al., 2020). Notably, bacteria such as Pseudomonas, Bacillus, Flavobacterium, Micrococcus sp. and Enterobacterhave been tested for their efficacy in handling a variety of heavy metals (Meliani & Bensoltane, 2016). Their robust biosorption ability is attributed to the substantial surface area ratios and potential active chemisorption sites, such as teichoic acid, on the cell wall (Mosa et al., 2016). Mixed cultures of microbes, thriving in biofilms, demonstrate enhanced stability and longevity (Sannasi et al., 2006). These biofilms serve as habitats for microbes. The microorganisms actively construct and engineer these biofilms, leading to the development of microbial populations (Flemming & Wingender, 2010). Biofilms are stable communities of mono- or multi-specific bacterial cells, with or without eukaryotes, attached to surfaces under various conditions (Costerton, 1995). This breakthrough in using microbes and their products for degrading and cleaning up a wide range of pollutants offers valuable benefits and should be integrated into protocols aimed at minimizing the environmental impact of corrosive releases without causing disruption to land or animal ecosystems.

## 5.7 Applications of Biofilm Linked Remediation

Various human activities, such as the production of natural gas, extraction of crude oil, and mining, have led to a significant rise in environmental pollution. Harmful and persistent pollutants, such as dyes, oils, pesticides, heavy metals, herbicides, phenolic compounds, and polyaromatic hydrocarbons (PAHs), are generated as byproducts of these activities. To address this issue, bacteria and biofilms emerge as promising solutions. Biofilms, consisting of microbes, prove beneficial for the bioremediation of micropollutants, as they can absorb or break down a wide range of harmful contaminants through various catabolic pathways (Edwards & Kjellerup, 2013).

### *5.7.1 Heavy Metal Detoxification*

Many heavy metals such as copper, zinc, nickel, cadmium, and cobalt have been effectively treated through the application of diverse biofilm reactors. In mining environments, biofilm-forming sulfate-reducing bacteria (SRB) prove particularly valuable for extracting metals from metal-contaminated water and converting them into precipitates of metal sulfides (Smith & Gadd, 2000; Muyzer & Stams, 2008). Recent research indicates that specific bacterial strains possess the ability to create electroactive films or electrochemically active biofilms (EAB). These biofilms directly engage in electron exchange with a conductive solid surface (Erable et al., 2009). EAB are increasingly being investigated in the realm of bioremediation and are employed to reduce heavy metals in contaminated groundwater and soil (Cong et al., 2013; Li et al., 2008). Numerous researchers have investigated and documented the biosorption of various heavy metals, including Cd(II), Cr(VI), Pb(II), Hg(II), Ni(II), Zn(II), Cu(II), and Mn(II), utilizing diverse types of bacterial biomass. The utilization of indigenous bacterial biomass, such as Arthrobacter sp. SUK 1205, Escherichia coli, Staphylococcus epidermidis, Leptothrixcholodnii SP-6SL, Pseudomonas sp. Lk9, Bacillus subtilis, Bacillus cereus, Shewanellaoneidensis, Penicillium simplicissimum, Stenotrophomonas spp., and Pseudomonas putida

CZ1, isolated from various metal-contaminated sites in bioremediation processes, appears to be effective. These bacterial strains have the potential to successfully eliminate heavy metals from contaminated waste, as demonstrated in studies by Luo et al. (2014), Dey and Paul (2015), Nell et al. (2016), Quiton et al. (2018), Pani et al. (2017), and Yu and Fein (2015).

In a study conducted by White and Gadd (2000), they investigated sulphate-reducing bacterial biofilms grown in continuous culture exposed to a medium containing 20–200 mM copper. The findings revealed the accumulation of copper in the form of copper sulfide within the biofilm. Simultaneously, an increase in the extracellular polymeric substance (EPS) content of the biofilm was observed, indicating the role of EPS and biofilms in trapping metal precipitates. Another study by Macaskie et al. (2000) observed that polycrystalline $NaUO_2PO_4$ accumulated in and around the cell wall of Citrobacter sp. N14, adsorbing to lipopolysaccharide and facilitating the bioprecipitation of the uranium salt. Additionally, Macaskie and Dean (1987) investigated the use of immobilized biofilm of Citrobacter sp. for the removal of uranium and lead from aqueous flow, with a significant portion of the exposed metals accumulating as insoluble metal phosphate.

In the context of Pd(II) uptake, Burkholderiacepacia biofilms were found to undergo simultaneous sorption and biomineralization processes, as reported by Templeton et al. (2003). Biofilms, characterized by their heterogeneity maintained by EPS, play a crucial role in these processes, preventing the production of toxic compounds during biosorption. The bacterial cell wall emerges as an eco-friendly means for adsorbing heavy metal particles due to its numerous anionic functional groups capable of binding to heavy metals, including peptidoglycan, teichoic acids, phospholipids, and lipopolysaccharides (Singh et al., 2021).

### 5.7.2 Oil Spill Remediation

Petroleum-related activities, encompassing production, storage, transportation, refining, and processing, often lead to the occurrence of petroleum hydrocarbon spills and discharges. These incidents result from blowout accidents during oilfield development, leakage from oil pipelines and storage tanks, accidents involving oil tankers and tankers, oil well waxing, and overhauls of refineries and petrochemical production equipment (Wang et al., 2018; Chen et al., 2015). Although treating oil pollution poses challenges, bacteria specialized in degrading petroleum hydrocarbons have naturally evolved due to their proximity to these compounds in the environment. The occurrence of oil spills can even influence the composition of hydrocarbon-degrading microbes in the soil (Song & Bartha, 1990). In marine environments, hydrocarbon-degrading bacteria with rapid growth capabilities can be employed for the decontamination of oil spills (Ron & Rosenberg, 2014).

In addressing petroleum degradation, various microorganisms such as Pseudomonas, Arthrobacter, Rhodococcus, Bacillus, Alcanivorax, and Cycloclasticus spp. of gamma proteobacteria have been utilized (Harayama et al.,

2004). Ongoing efforts involve the continuous screening of new microbial strains capable of biofilm-mediated oil degradation (Dasgupta et al., 2013). In the case of petroleum degradation, microbial consortia that include Bacillus subtilis and Acinetobacter radioresistens, along with a surfactant-producing strain, have demonstrated superior degradation capabilities compared to consortia consisting solely of degraders (Mnif et al., 2015). A field study revealed that bioaugmentation with an artificial consortium containing Aeromonas hydrophila, Alcaligenes xylosoxidans, Gordonia sp., Pseudomonas fluorescens, Pseudomonas putida, Rhodococcusequi, S. maltophilia, and Xanthomonas sp. significantly contributed to high biodegradation efficiency (89%) during a 365-day treatment of diesel oil-contaminated soil (Szulc et al., 2014). This underscores the potential of tailored microbial communities in enhancing the efficiency of bioremediation processes for petroleum-contaminated environments.

### 5.7.3 *Wastewater Treatment*

The electronics, pharmaceuticals, and cosmetics industries, along with various inorganic and organic pollutants, contribute to the release of hazardous micropollutants into the environment, leading to water pollution. Addressing this issue, bacteria and biofilms emerge as promising solutions. Microbes within biofilms offer significant benefits for micropollutant bioremediation, as they can absorb or break down a wide range of severe contaminants through various catabolic pathways (Edwards & Kjellerup, 2013).In a study by Li et al. (2015), a combination of three biofilm-forming bacterial strains, namely B. subtilis N4, E2, and E3, was utilized along with an acetonitrile-degrading bacterium (Rhodococcusrhodochrous BX2) to mitigate acetonitrile toxicity. This approach demonstrates the potential of employing biofilm-forming bacterial consortia for the reduction of specific micropollutants, showcasing the versatility of microbial solutions in addressing environmental pollution from industrial sources.

In a related study, a recombinant bacterium capable of producing biofilms (B. subtilis N4-pHT01-nit) was found to exhibit high efficacy in degrading elevated concentrations of acetonitrile (800 mg L_1) within the following 24 h, especially when utilized in conjunction with Moving Bed Biofilm Reactor (MBBR) technology (Maurya & Raj, 2020). This highlights the potential of engineered bacteria with biofilm-forming capabilities in efficiently degrading pollutants, showcasing a valuable approach for environmental remediation. The significance of biofilms in the degradation of diverse organic and inorganic contaminants is crucial for bioremediation and wastewater treatment processes. Sulfate-reducing bacteria (SRB) form a major microbial community involved in biofilm formation, particularly in metal-contaminated waters. SRBs are highly efficient in anaerobically degrading various organic pollutants and precipitating heavy metals from wastewater as metal sulfide (Top et al., 1998). This underscores the multifaceted role of biofilms and specific microbial communities in addressing pollution challenges across different environmental contexts.

Costley and Wallis (2001) observed an impressive 84% removal of Cd2+, Cu2+, and Zn2+ from synthetic wastewater using a rotating biological contactor with alternating sorption and desorption cycles. In another study, the applicability and feasibility of treating acidic sulfate, zinc, and iron-containing wastewater with sulfate-reducing bacteria (SRB) suspension and biofilm were investigated. The addition of lactate to the wastewater enriched sulfate-reducing bacteria in both suspension and biofilms, leading to the precipitation and accumulation of metals in the biofilm as sulfides (Kaksonen et al., 2003). Titanium dioxide nanoparticles, commonly found in cosmetics, paints, and various household items, pose a potential environmental hazard when discharged into wastewater from residential areas globally. Due to the photocatalytic activation of the titania element, these nanoparticles can have adverse effects. To mitigate the toxicity of nanoparticles, Mathur et al. (2017) studied the biofilm-mediated absorption and removal of titanium dioxide particles, facilitated by the presence of humic and surfactant compounds in the wastewater. Biofilms are increasingly utilized as early warning systems or indicators for monitoring and evaluating heavy metal contamination in rivers and streams. The rapid structural and physiological alterations of biofilms in the presence of toxicants make them valuable for assessing environmental pollution. Choosing biofilms as an indicator system offers benefits such as pollutant adsorption, rapid development, and easy sampling methods (Fuchs et al., 1997).

## 5.8 Challenges of Biofilm Linked Remediation

Biofilm-linked remediation, despite its potential for efficient environmental cleanup, faces several challenges that need careful consideration. These challenges span various stages of the remediation process and must be addressed to optimize the effectiveness of biofilm-mediated strategies.

### *5.8.1 Matrix Complexity*

Biofilm matrices are intricate structures composed of extracellular polymeric substances (EPS) that create a complex environment. This complexity can hinder the diffusion of nutrients, oxygen, and other remediation agents, affecting their effectiveness in reaching microbial cells. The thickness of biofilms can vary, and the penetration of remediation agents may be limited in thicker biofilms. This can reduce the efficacy of the treatment in reaching the inner layers where microbial activity may be crucial. The thickness and cell density of a biofilm are significant variables to consider while assessing it (Mathur et al., 2017). Another crucial factor is the matrix porosity, which influences the biofilm's ability and structure. According to study, thin biofilms contain fewer holes and a greater cell density than thick biofilms. Biofilm may use these perforations as a solid–liquid partitioning substrate. Consequently, the biofilm's ability to break down phenol is related to its thickness; therefore, thickness is suggested for usage as an effective bioremediation strategy (Crampon et al., 2018). Biofilms exhibit microheterogeneity due to the presence of various microenvironments within the matrix. This diversity can result in uneven distribution of microbial activity, leading to localized remediation effectiveness and potential recontamination of treated areas.

### 5.8.2 *Operational Flexibility*

It is hindered by the sensitivity of biofilms to environmental conditions. Changes in factors such as temperature, pH, or nutrient availability can alter the biofilm's structure and function, affecting its remediation capabilities. Achieving optimal conditions for biofilm growth and activity requires continuous monitoring and adjustment, demanding a high level of operational vigilance. Moreover, the scalability of biofilm-linked remediation poses a challenge (Sharma et al., 2020). The transition from controlled environments to diverse and unpredictable field conditions introduces uncertainties and operational hurdles. The management of biofilm-associated risks is another operational concern. While biofilms are harnessed for their remediation potential, there is a need to address potential unintended consequences. For instance, biofilms might inadvertently contribute to the release of certain pollutants or create byproducts that require further management. Additionally, the implementation of biofilm-linked remediation on contaminated sites with complex matrices demands tailored strategies. The heterogeneity of environmental matrices presents difficulties in achieving uniform biofilm distribution and activity. Site-specific conditions, such as the presence of competing microorganisms or varying substrate availability, need to be carefully considered in operational planning. Overcoming these challenges requires a thorough understanding of biofilm dynamics, continuous monitoring, and the development of strategies that balance the benefits of biofilm-mediated remediation with the complexities of real-world environmental conditions (Singh et al., 2006).

### 5.8.3 *Quorum Sensing*

Quorum sensing is a sophisticated communication mechanism employed by bacteria to coordinate group behavior through the release and detection of signaling molecules. While quorum sensing is essential for the cohesive and organized development of biofilms, it simultaneously presents challenges, particularly in the context of biofilm resilience and resistance to external interventions (Parsek and Greenberg 2005). The signaling molecules involved in quorum sensing, often referred to as autoinducers, play a dual role in biofilm dynamics. On one hand, they act as messengers that facilitate the synchronization of bacterial activities, promoting the formation of structured and functional biofilms. On the other hand, these molecules contribute to the robustness of biofilms, rendering them less susceptible to external disturbances and remediation strategies. The resilience of biofilms, fostered by quorum sensing, arises from the coordinated response of bacterial cells to the changing environment. When a quorum, or a critical cell density, is achieved through the accumulation of signaling molecules, the bacterial community becomes more synchronized and can collectively adapt to stressors. This synchronization enables the biofilm to resist external interventions that might disrupt its structural integrity or target the embedded microbial populations (Hall-Stoodley et al., 2004).

One of the challenges posed by quorum sensing is the potential enhancement of biofilm resistance to antimicrobial agents, disinfectants, or other remediation efforts. As the signaling molecules facilitate communication among bacteria, they contribute to the activation of defense mechanisms within the biofilm. This can include the upregulation of genes associated with antibiotic resistance and the production of protective EPS layers, making it more difficult for external agents to penetrate and eliminate the biofilm.The intricate interplay between quorum sensing and biofilm resilience underscores the complexity of microbial communities and their adaptive strategies. Understanding these dynamics is crucial for the development of effective biofilm remediation strategies.

Bioremediation can face challenges when essential nutrients necessary for supporting microbial growth are limited, especially in situations of high pollution levels. The efficiency of bioremediation is also influenced by the bioavailability of pollutants to microbes, with pollutants not enclosed by materials like clay being more conducive to biodegradation. Successful biofilm-mediated bioremediation demands adequate technical expertise and an interdisciplinary approach involving fields such as environmental microbiology, civil engineering, and soil science. In cases where rapid microbial growth occurs and reactor capacity relies on oxygen diffusion, biofilm reactors may not be suitable. These limitations highlight the importance of assessing the effectiveness of bioremediation relative to other available methods for environmental pollutant decontamination (Heitzer & Sayler, 1993). Despite advances, there is still much to learn about the complexities of biofilm behavior, especially in real-world, dynamic environments. Insufficient understanding may hinder the development of targeted and efficient remediation strategies. Addressing these challenges requires ongoing research and innovation to improve the understanding of biofilm dynamics and enhance the efficiency of biofilm-linked remediation processes.

## 5.9 Future Directions in Biofilm-Linked Bioremediation

The future direction in biofilm-linked remediation is likely to focus on several key areas to enhance efficiency, sustainability and applicability. Continued research and development will likely lead to the creation of advanced bioremediation technologies that leverage biofilms. This could involve the engineering of biofilms with enhanced pollutant-degrading capabilities, increased resistance to environmental stressors, and optimized metabolic pathways. The field may witness more emphasis on genetic engineering to tailor biofilms for specific remediation purposes. This could involve modifying microbial strains within the biofilm to express genes that enhance pollutant degradation or to improve the overall resilience and adaptability of the biofilm.

To expedite the processes of biofilm-mediated bioremediation, comprehensive approaches such as multi-omics platforms and gene-centric strategies are essential. These methodologies aim to unravel the intricate mechanisms governing the interaction between metals and biofilms, as well as the strategies employed for metal sequestration. Moreover, the application of genetic engineering holds potential in crafting highly efficient genetically engineered biofilm communities (Sharma, 2022). This genetic engineering approach can also be harnessed to fabricate microbial electroactive biofilms endowed with superior extracellular transport capabilities, suitable for integration into bioelectrochemical systems (Jasu & Ray, 2021). A groundbreaking approach to enhance bioremediation harnesses the inherent transformation mechanisms within biofilms. This process involves the assimilation of DNA containing catabolic genes that facilitate the breakdown of specific contaminants. Engineered microorganisms, known as genetically modified microbes (GEMs), have been created to metabolize a diverse range of pollutants. The horizontal transfer of genes responsible for biodegradation from GEMs to members of biofilm communities presents an opportunity to further enhance the efficiency of the biodegradation process. Additionally, by cloning genes related to biosurfactant synthesis and leveraging the chemotactic capabilities of GEMs, the biodegradative potential of these modified microorganisms could be further optimized.

To enhance the resistance of microorganisms against diverse heavy metal ions, a viable strategy involves the overexpression of functional genes. This genetic manipulation assists microbes in overcoming the challenges posed by complex heavy metal toxicity (Yin et al., 2019). Future research endeavors should concentrate on identifying these functional genes within the microbial diversity, emphasizing those that support growth and metabolism. The ultimate goal is to develop commercial bioinoculants tailored for effective metal remediation (Jasu & Ray, 2021). Integration of synthetic biology approaches could play a significant role in designing biofilms with custom functionalities. This may involve the construction of synthetic microbial consortia within biofilms, allowing for precise control over their composition and performance in remediation processes. Further, nanotechnology may be integrated into biofilm-linked remediation for improved pollutant capture and degradation. Nano-sized materials could be designed to enhance the catalytic activity of biofilms or to provide additional functionalities, such as targeted pollutant binding. The development of biofilms with sensing capabilities could also revolutionize remediation efforts. Smart biofilms capable of detecting pollutant levels, environmental conditions, and the overall health of the remediation process could allow for real-time monitoring and adaptive responses.

## 5.10 Conclusion

In challenging polluted environments, many bacteria demonstrate a remarkable ability to form biofilms, contributing to their enhanced survival. The intricate structure of the extracellular polymeric substance (EPS) matrix, coupled with features like intercellular gene transfer, quorum sensing, and chemotaxis, positions these biofilm-forming bacteria as promising candidates for bioremediation efforts targeting various toxic and persistent compounds. Biofilms offer an appealing tool for monitoring and evaluating environmental contaminants. Bacterial biofilms, comprising single or multiple species from diverse habitats, prove effective in both in situ and ex situ bioremediation processes. Enhancing our understanding of microbial processes in pollution tolerance and degradation can significantly benefit bioremediation approaches. Exploring the mechanisms and genes involved in biofilm formation opens avenues for innovative bioremediation methods. The study of efficient gene transfer to develop genetically-engineered biofilms and the thorough exploration of bioremediation across aquatic and terrestrial systems hold the potential to establish sustainable green technologies in pollution control. This review delves into the comprehensive understanding of biofilm composition, development, mechanisms, and the application of biofilms in the bioremediation of heavy metal contaminants, contributing valuable insights to the field.

## References


Allegrucci, M., et al. (2006). Phenotypic characterization of Streptococcus pneumoniae biofilm development. *Journal of Bacteriology, 188*(6), 2325–2335.

Bielik, B., Molnár, L., Vrabec, V., Andrášiová, R., Maruščáková, I. C., Nemcová, R., Toporčák, J., & Mudroňová, D. (2021). Acta Veterinaria Hungarica. https://doi.org/10.1556/004.2020.00057

Caruso, C., et al. (2018). Production and biotechnological potential of extracellular polymeric substances from sponge-associated Antarctic bacteria. *Applied and Environmental Microbiology, 84*(4), e01624–e01617.

Chen, M., Xu, P., Zeng, G., Yang, C., Huang, D., & Zhang, J. (2015). Bioremediation of soils contaminated with polycyclic aromatic hydrocarbons, petroleum, pesticides, chlorophenols and heavy metals by composting: Applications, microbes and future research needs. *Biotechnology Advances, 33*, 745–755. https://doi.org/10.1016/j.biotechadv.2015.05.003

Cong, Y., Xu, Q., Feng, H., et al. (2013). Efficient electrochemically active biofilm denitrification and bacteria consortium analysis. *Bioresource Technology, 132*, 24–27.

Costerton, J. W. (1995). Overview of microbial biofilms. *Journal of Industrial Microbiology and Biotechnology, 15*(3), 137–140.

Costley, S. C., & Wallis, F. M. (2001). Bioremediation of heavy metals in a synthetic wastewater using a rotating biological contactor. *Water Research, 35*, 3715–3723.

Crampon, M., et al. (2018). Do natural biofilm impact nZVI mobility and interactions with porous media? A column study. *Science of the Total Environment, 610*, 709–719.

Das, N., Basak, L. V. G., & Salam, J. A. (2017). Application of biofilms on remediation of pollutants—An overview. *Journal of Microbiolog and Biotechnology, 2*(5), 783–790.

Dasgupta, D., Ghosh, R., & Sengupta, T. K. (2013). Biofilm-mediated enhanced crude oil degradation by newly isolated pseudomonas species. *ISRN Biotechnology, 2013*, 250749.

Decho, A. W. (2000). Microbial biofilms in intertidal systems: An overview. *Continental Shelf Research, 20*, 1257–1273.

Delangiz, N., Varjovi, M. B., Lajayer, B. A., & Ghorbanpour, M. (2020). Beneficial microorganisms in the remediation of heavy metals. In *Molecular aspects of plant beneficial microbes in agriculture*. Elsevier. https://doi.org/10.1016/b978-0-12-818469-1.00034-1

Dey, S., & Paul, D. K. (2015). Hexavalent chromate reduction during growth and by immobilized cells of Arthrobacter sp. SUK 1205. *Science, 34*(3), 158–168.

Di Donato, P., et al. (2016). Recent advances in the study of marine microbial biofilm: From the involvement of quorum sensing in its production up to biotechnological application of the polysaccharide fractions. *Journal of Marine Science and Engineering, 4*(2), 34.

Edwards, S. J., & Kjellerup, B. V. (2013). Applications of biofilms in bioremediation and biotransformation of persistent organic pollutants, pharmaceuticals/personal care products, and heavy metals. *Applied Microbiology and Biotechnology, 97*(23), 9909–9921. https://doi.org/10.1007/s00253-013-5216-z

Erable, B., Duţeanu, N. M., Ghangrekar, M. M., et al. (2009). Application of electro-active biofilms. *Biofouling, 26*, 57–71.

Fernández Ramírez, M. D., Smid, E. J., Abee, T., & Nierop Groot, M. N. (2015). Characterisation of biofilms formed by Lactobacillus plantarum WCFS1 and food spoilage isolates. *International Journal of Food Microbiology, 207*, 23–29. https://doi.org/10.1016/j.ijfoodmicro.2015.04.030

Flemming, H. C. (1995). Sorption sites in biofilms. *Water Science and Technology, 32*(7), 27–33.

Flemming, H. C., & Wingender, J. (2010). The biofilm matrix. *Nature Reviews Microbiology, 8*(9), 623–633. https://doi.org/10.1038/nrmicro2415

Flemming, H.-C., Neu, T. R., & Wingender, J. (Eds.). (2016). The Perfect Slime: Microbial Extracellular Polymeric Substances (EPS) (Vol. 15). IWA Publishing. https://doi.org/10.2166/9781780407425

Fuchs, S., Haritopoulou, T., Schäfer, M., et al. (1997). Heavy metals in freshwater ecosystems introduced by urban rainwater runoff—Monitoring of suspended solids, river sediments, and biofilms. *Water Science and Technology, 36*, 277–282.

Fux, C. A. (2005). Survival strategies of infectious biofilms. *Trends in Biotechnology, 13*, 34–40.

Gentry, T., Rensing, C., & Pepper, I. A. N. (2004). New approaches for bioaugmentation as a remediation technology. *Critical Reviews in Environmental Science and Technology, 34*(5), 447–494.

Gupta, P., & Diwan, B. (2017). Bacterial exopolysaccharide mediated heavy metal removal: A review on biosynthesis, mechanism and remediation strategies. *Biotechnology Reports, 13*, 58–71. https://doi.org/10.1016/j.btre.2016.12.006

Hall-Stoodley, L., Costerton, J. W., & Stoodley, P. (2004). Bacterial biofilms: From the natural environment to infectious diseases. *Nature Reviews Microbiology, 2*(2), 95–108.

Harayama, S., Kasai, Y., & Hara, A. (2004). Microbial communities in oil-contaminated seawater. *Current Opinion in Biotechnology, 15*, 205–214.

Heitzer, A., & Sayler, G. S. (1993). Monitoring the efficacy of bioremediation. *Trends in Biotechnology, 11*(8), 334–343.

Horn, H., & Morgenroth, E. (2006). Transport of oxygen, sodium chloride, and sodium nitrate in biofilms. *Chemical Engineering Science, 61*, 1347–1356.

Jasu, A., & Ray, R. R. (2021). Biofilm mediated strategies to mitigate heavy metal pollution: A critical review in metal bioremediation. *Biocatalysis and Agricultural Biotechnology, 37*(September), 102183. https://doi.org/10.1016/j.bcab.2021.102183

Jamal, M., Ahmad, W., Andleeb, S., Jalil, F., Imran, M., Nawaz, M. A., Hussain, T., Ali, M., Rafiq, M., & Kamil, M. A. (2018). Bacterial biofilm and associated infections. *Journal of Chinese Medicine, 81*(1), 7–11.

Kaksonen, A. H., Riekkola-Vanhanen, M. L., & Puhakka, J. A. (2003). Optimization of metal sulfide precipitation in fluidized-bed treatment of acidic wastewater. *Water Research, 37*, 255–266.

Kumar, J., Dewulf, T.V.D, Wiele, & H.V., Langenhove. (2009). Bacterial dynamics of biofilm development during toluene degradation by Burkholderia vietnamiensis G4 in a gas phase membrane bioreactor. *Biotechnol, 19*, 1028–1033.

Kjelleberg, S., & Molin, S. (2002). Is there a role for quorum sensing signals in bacterial biofilms? *Current Opinion in Microbiology, 5*(3), 254–258.

Li, Z., Zhang, X., & Lei, L. (2008). Electricity production during the treatment of real electroplating wastewater containing Cr6+ using microbial fuel cell. *Process Biochemistry, 43*, 1352–1358.

Li, T., Lin, G., Podola, B., & Melkonian, M. (2015). Continuous removal of zinc from wastewater and mine dump leachate by a microalgal biofilm PSBR. *Journal of Hazardous Materials, 297*, 112–118.

Liu, P. W. G., Chang, T. C., Whang, L. M., Kao, C. H., Pan, P. T., & Cheng, S. S. (2011). Bioremediation of petroleum hydrocarbon contaminated soil: Effects of strategies and microbial community shift. *International Biodeterioration and Biodegradation, 65*(8), 1119–1127.

López, D., Vlamakis, H., & Kolter, R. (2010). Biofilms. *Cold Spring Harbor perspectives in biology, 2*(7), a000398. https://doi.org/10.1101/cshperspect.a000398

Luo, S., Li, X., Chen, L., Chen, J., Wan, Y., & Liu, C. (2014). Layer-by-layer strategy for adsorption capacity fattening of endophytic bacterial biomass for highly effective removal of heavy metals. *Chemical Engineering Journal, 239*, 312–321.

Macaskie, L. E., & Dean, A. C. R. (1987). Use of immobilized biofilm of Citrobacter sp. for the removal of uranium and lead from aqueous flows. *Enzyme and Microbial Technology, 9*, 2–4.

Macaskie, L. E., et al. (2000). Enzymatically mediated bioprecipitation of uranium by Citrobacter sp.: A concerted role for exocellular lipopolysaccharide and associated phosphatase in biomineral formation. *Microbiology, 146*, 1855–1867.

Mangwani, N., Kumari, S., & Das, S. (2017). Marine bacterial biofilms in bioremediation of polycyclic aromatic hydrocarbons (PAHs) under terrestrial condition in a soil microcosm. *Pedosphere, 27*(3), 548–558. https://doi.org/10.1016/S1002-0160(17)60350-3

Mathur, A., Parashar, A., Chandrasekaran, N., & Mukherjee, A. (2017). Nano-TiO2 enhances biofilm formation in a bacterial isolate from activated sludge of a wastewater treatment plant. *International Biodeterioration and Biodegradation, 116*, 17–25.

Maurya, A., & Raj, A. (2020). Recent advances in the application of biofilm in bioremediation of industrial wastewater and organic pollutants. In *Microorganisms for sustainability: Environmental health* (pp. 81–118). Elsevier.

Medfu Tarekegn, M., Zewdu Salilih, F., & Ishetu, A. I. (2020). Microbes used as a tool for bioremediation of heavy metal from the environment. *Cogent Food and Agriculture, 6*(1). https://doi.org/10.1080/23311932.2020.1783174

Meliani, A., & Bensoltane, A. (2016). Biofilm-mediated heavy metals bioremediation in PGPR pseudomonas. *Journal of Bioremediation & Biodegradation, 7*(370), 2.

Miller, M. B., & Bassler, B. L. (2001). Quorum sensing in bacteria. *Annual Review of Microbiology, 55*(1), 165–199.

Mitra, A., & Mukhopadhyay, S. (2015). Biofilm-mediated decontamination of pollutants from the environment. *AIMS Bioengineering, 3*(1), 44–59.

Mitra, S., Sana, B., & Mukherjee, J. (2014a). Ecological roles and biotechnological applications of marine and intertidal microbial biofilms. *Advances in Biochemical Engineering/Biotechnology, 146*, 163–205. https://doi.org/10.1007/10_2014_271

Mitra, S., Sana, B., & Mukherjee, J. (2014b). Ecological roles and biotechnological applications of marine and intertidal microbial biofilms. In *Productive biofilms* (pp. 163–205). Springer.

Mnif, I., Mnif, S., Sahnoun, R., et al. (2015). Biodegradation of diesel oil by a novel microbial consortium: Comparison between co-inoculation with biosurfactant-producing strain and exogenously added biosurfactants. *Environmental Science and Pollution Research International, 22*, 14852–14861.

Mohapatra, R. K., Parhi, P. K., Patra, J. K., Panda, C. R., & Thatoi, H. N. (2017). Biodetoxification of toxic heavy metals by marine metal resistant bacteria—A novel approach for bioremediation of the polluted saline environment. In *Microbial biotechnology* (pp. 343–376). Springer.

Mohapatra, R. K., Parhi, P. K., Patra, J. K., Panda, C. R., & Thatoi, H. N. (2018). Biodetoxification of toxic heavy metals by marine metal resistant bacteria—A novel approach for bioremediation of the polluted saline environment. *Microbial Biotechnology, 1*. https://doi.org/10.1007/978-981-10-6847-8_15

Mohapatra, R. K., Behera, S. S., Patra, J. K., Thatoi, H., & Parhi, P. K. (2019). Potential application of bacterial biofilm for bioremediation of toxic heavy metals and dye-contaminated environments. In *New and future developments in microbial biotechnology and bioengineering*. Elsevier. https://doi.org/10.1016/B978-0-444-64279-0.00017-7

Mosa, K. A., Saadoun, I., Kumar, K., Helmy, M., & Dhankher, O. P. (2016). Potential biotechnological strategies for the cleanup of heavy metals and metalloids. *Frontiers in Plant Science, 7*, 303.

Molin, S. & Tolker-Nielsen, T. (2003). Gene transfer occurs with enhanced efficiency in biofilms and induces enhanced stabilization of the biofilms structure. *Current Opinion in Biotechnology, 14*, 255–261.

Muyzer, G., & Stams, A. J. (2008). The ecology and biotechnology of sulfate-reducing bacteria. *Nature Reviews Microbiology, 6*, 441–454.

Nell, R. M., Szymanowski, J. E., & Fein, J. B. (2016). Divalent metal cation adsorption onto Leptothrixcholodnii SP-6SL bacterial cells. *Chemical Geology, 439*, 132–138.

Pani, T., Das, A., & Osborne, J. W. (2017). Bioremoval of zinc and manganese by bacterial biofilm: A bioreactor-based approach. *Journal of Photochemistry and Photobiology B: Biology, 175*, 211–218.

Parsek, M. R., & Greenberg, E. P. (2005). Sociomicrobiology: The connections between quorum sensing and biofilms. *Trends in Microbiology, 13*(1), 27–33.

Paul, D., et al. (2005). Accessing microbial diversity for bioremediation and environmental restoration. *Trends in Biotechnology, 23*(3), 135–142.

Quiton, K. G., Doma, J. B., Futalan, C. M., & Wan, M. W. (2018). Removal of chromium (VI) and zinc (II) from aqueous solution using kaolin-supported bacterial biofilms of Gram-negative E. coli and gram-positive Staphylococcus epidermidis. *Sustainable Environment Research, 28*(5), 206–213.

Rastegari, A. A., Yadav, A. N., & Yadav. (2020). *New and future developments in microbial biotechnology and bioengineering: Trends of microbial biotechnology for sustainable agriculture and biomedicine systems: Diversity and functional perspectives*. Elsevier.

Rashtchi, P., Tempelaars, M., van der Linden, E., Abee, T., & Habibi, M. (2022). Lactobacillus plantarum strains show diversity in biofilm formation under flow conditions. *Heliyon, 8*(12), e12602. https://doi.org/10.1016/j.heliyon.2022.e12602

Ron, E. Z., & Rosenberg, E. (2014). Enhanced bioremediation of oil spills in the sea. *Current Opinion in Biotechnology, 27*, 191–194.

Sannasi, P., Kader, J., Othman, O., & Salmijah, S. (2006). Single and multi-metal removal by an environmental mixed bacterial isolate. In *Modern multidisciplinary applied microbiology: Exploiting microbes and their interactions* (pp. 136–141).

Sharma, P. (2022). Role and significance of biofilm-forming microbes in phytoremediation—A review. *Environmental Technology and Innovation, 25*, 102182. https://doi.org/10.1016/j.eti.2021.102182

Sharma, A., Jamali, H., Vaishnav, A., Giri, B. S., & Srivastava, A. K. (2019). Microbial biofilm: An advanced eco friendly approach for bioremediation. In *New and future developments in microbial biotechnology and bioengineering*. Elsevier. https://doi.org/10.1016/B978-0-444-64279-0.00015-3

Sharma, A., Jamali, H., Vaishnav, A., Giri, B. S., & Srivastava, A. K. (2020). Microbial biofilm: An advanced eco-friendly approach for bioremediation. In *New and future developments in microbial biotechnology and bioengineering: Microbial biofilms* (pp. 205–219). Elsevier.

Shukla, S. K., Mangwani, N., Rao, T. S., & Das, S. (2014). *Biofilm-mediated bioremediation of polycyclic aromatic hydrocarbons*. Elsevier Inc. https://doi.org/10.1016/B978-0-12-800021-2.00008-X

Shukla, S. K., Mangwani, N., Karley, D., & Rao, T. S. (2017). Bacterial biofilms and genetic regulation for metal detoxification. In *Handbook of metal-microbe interactions and bioremediation* (pp. 317–332). CRC Press. https://doi.org/10.1201/9781315153353-22

Singh, R., Paul, D., & Jain, R. K. (2006). Biofilms: Implications in bioremediation. *Trends in Microbiology, 14*(9), 389–397.

Singh, A., Saini, P., Srivastava, U., Ahmed, M., & Iqbal, U. (2021). Cadmium toxicology and its prevention through lactic acid bacteria. *Journal of Food and Agriculture Research, 1*(2), 221–248.

Sinha, S., et al. (2018). Removal of Congo red dye from aqueous solution using Amberlite IRA-400 in batch and fixed bed reactors. *Chemical Engineering Communications, 205*(4), 432–444. https://doi.org/10.1080/00986445.2017.13993366

Smith, W. L., & Gadd, G. M. (2000). Reduction and precipitation of chromate by mixed culture sulfate-reducing bacterial biofilms. *Journal of Applied Microbiology, 88*, 983–991.

Song, H. G., & Bartha, R. (1990). Effects of jet fuel spills on the microbial community of soil. *Applied and Environmental Microbiology, 56*, 646–651.

Stanley, N. R., & Lazazzera, B. A. (2004). Environmental signals and regulatory pathways that influence biofilm formation. *Molecular Microbiology, 52*(3), 917–924.

Stanley, N. R., et al. (2003). Identification of catabolite repression as a physiological regulator of biofilm formation by Bacillus subtilis by use of DNA microarrays. *Journal of Bacteriology, 185*(6), 1951–1957.

Sutherland, I. W. (2001). The biofilm matrix—An immobilized but dynamic microbial environment. *Trends in Microbiology, 9*(5), 222–227.

Szulc, A., Ambro˙zewicz, D., Sydow, M., Ławniczak, Ł., Piotrowska-Cyplik, A., Marecik, R., et al. (2014). The influence of bioaugmentation and biosurfactant addition on bioremediation efficiency of diesel-oil contaminated soil: Feasibility during field studies. *Journal of Environmental Management, 132*, 121–128. https://doi.org/10.1016/j.jenvman.2013.11.006

Templeton, A. S., Trainor, T. P., Spormann, A. M., Newville, M., Sutton, S. R., & Dohnalkova, A. (2003). Sorption versus biomineralization of Pd(II) within Burkholderiacepacia biofilms. *Environmental Science & Technology, 37*, 300–307.

Top, E. M., et al. (1998). Enhancement of 2,4-dichlorophenoxyacetic acid (2,4-D) degradation in soil by dissemination of catabolic plasmids. *Antonie Van Leeuwenhoek, 73*, 87–94.

Tyagi, M., da Fonseca, M. M. R., & de Carvalho, C. C. (2011). Bioaugmentation and biostimulation strategies to improve the effectiveness of bioremediation processes. *Biodegradation, 22*(2), 231–241.

Vaishnav, A., Kumari, S., Jain, S., Varma, A., Tuteja, N., & Choudhary, D. K. (2016). PGPR-mediated expression of salt tolerance gene in soybean through volatiles under sodium nitroprusside. *Journal of Basic Microbiology, 56*(11), 1274–1288. https://doi.org/10.1002/jobm.201600188

Wang, C., Liu, X., Guo, J., Lv, Y., & Li, Y. (2018). Biodegradation of marine oil spill residues using aboriginal bacterial consortium based on Penglai 19-3 oil spill accident in China. *Ecotoxicology and Environmental Safety, 159*, 20–27. https://doi.org/10.1016/j.ecoenv.2018.04.059

Watnick, P. I., & Kolter, R. (1999). Steps in the development of a vibrio cholerae El tor biofilm. *Molecular Microbiology, 34*(4), 586–595.

White, C., & Gadd, G. M. (2000). Copper accumulation by sulfate-reducing bacterial biofilms. *FEMS Microbiology Letters, 183*, 313–318.

Whitfield, C., & Valvano, M. A. (1993). Biosynthesis and expression of cell-surface polysaccharides in Gram-negative bacteria. In A. H. Rose (Ed.), Advances in Microbial Physiology (Vol. 35, pp. 135–246). Academic Press. https://doi.org/10.1016/S0065-2911(08)60099-5

Yin, K., Wang, Q., Lv, M., & Chen, L. (2019). Microorganism remediation strategies towards heavy metals. *Chemical Engineering Journal, 360*(October), 1553–1563. https://doi.org/10.1016/j.cej.2018.10.226

Yu, Q., & Fein, J. B. (2015). The effect of metal loading on Cd adsorption onto Shewanellaoneidensis bacterial cell envelopes: The role of sulfhydryl sites. *Geochimica et Cosmochimica Acta, 167*, 1–10.

# Chapter 6
# Microbial Bioprospecting for Biorefinery Application: An Overview of a Sustainable and Renewable Source of Energy

**Contents**



 
P. Saini, P. Mishra, *Bioprospecting of Microbial Resources for Agriculture, Environment and Bio-chemical Industry*, https://doi.org/10.1007/978-3-031-63844-2_6

## 6.1 Introduction

Energy plays a pivotal role in discussions concerning sustainable development across economic, social, and environmental dimensions (Midilli et al., 2006; Dincer, 1999). Fossil-based energy, encompassing resources like coal, petroleum, and natural gas, is distinct from green energy. It's a well-established fact that fossil energy sources are non-renewable. To understand the necessity for sustainable energy strategies, particularly in the context of green energy, it's essential to first address the adverse impacts associated with fossil fuels.

The extensive use of fossil fuels in both industrial and non-industrial sectors has resulted in significant human health and welfare issues, as extensively documented (Barreto et al., 2003; Bockris, 2003). The root cause of these problems lies in the pervasive utilization of fossil-based technologies and strategies by humans throughout history to govern societies, nations, and the world. This situation has reached a point of intolerability. Urgent action is required to develop green energy strategies that promise a sustainable future devoid of detrimental environmental and societal consequences. So, what is green energy? It can be defined as an energy source with minimal to zero environmental impact, more environmentally benign and sustainable, derived from sources like solar, hydro, biomass, wind, geothermal, and others. Green energy significantly mitigates the negative effects of fossil energy resources, reduces emissions from electricity generation, curtails greenhouse gases, supports environmental enhancement, and fulfills the demand for clean energy in industrial and non-industrial applications.

Given the manifold benefits of green energy, its sustainability in supply and progress emerges as a critical element in the interplay between nature and society. Sustainable development necessitates energy resources that are readily available at a reasonable cost and impose little to no negative societal impacts. Evidently, fossil fuels and their finite nature do not align with the requisites of sustainability, while green energy sources exhibit long-term sustainability (Dincer & Rosen, 2004). Notably, affordable green energy is a fundamental driver for sustainable technological advancement, industrial productivity, and improved living standards. Consequently, it's imperative to establish enduring and effective strategies for sustainable green energy utilization (Midilli et al., 2006).

As reported by the U.S. Energy Information Administration (EIA, 2017), renewable energy sources are continually replenished, in contrast to finite fossil fuels. The five commonly used renewable energy sources worldwide, as noted by EIA, include biofuels (comprising biomass, ethanol, and biodiesel), hydropower, geothermal, wind, and solar. Biomass stands out as a significant renewable energy source utilized in the United States and globally.

Microorganisms are omnipresent, displaying resilience in extreme conditions. The diversity among microorganisms is extensive, with only 1% of total biodiversity explored to date due to limitations in laboratory culturing techniques (Akondi & Lakshmi, 2013). Microorganisms exhibit a wide range of capabilities, from causing diseases to yielding valuable products such as antibiotics, enzymes, and bioactive compounds. They also contribute to bioremediation and the degradation of organic waste materials. Microorganisms play diverse roles in ecosystems, offering novel and valuable products that benefit society and contribute to environmental remediation. Their rapid growth rates, ease of isolation, and efficient product extraction methods make them well-suited for these applications. Consequently, microbial bioprospecting assumes a crucial role in advancing sustainability. In light of the aforementioned discussion, this study aims to shed light on the concepts of biorefineries, the demand for energy, the role of microbes in biorefineries, and the production of biofuels through microbial processes.

## 6.2 Biorefinery Concept and Energy Demand

Biorefineries, akin to petroleum refineries, utilize biomass instead of crude oil to generate transportation fuels, heat, power, chemicals, and materials. The carbon-neutral and renewable nature of biomass led to its recognition as a potential solution to mitigate climate change threats and meet the ever-growing demands for energy, fuels, chemicals, and materials at the World Economic Forum in 2008 and 2009 (World Economic Forum, 2010). Currently, various definitions of biorefineries exist, shaped by the type of activities and stakeholders involved. For example, one definition describes a biorefinery as a facility that integrates biomass conversion processes and equipment to produce fuels, power, and value-added chemicals from biomass (van Ree & Annevelink, 2007). Another definition characterizes it as a refining facility where biomass is converted into fuel, chemicals, materials, and other products all within the same plant. Europabio (2007) defines a biorefinery as an integrated cluster of bio-industries utilizing various technologies to produce chemicals, biofuels, food ingredients, and power from raw biomass materials (Europabio, 2007). Presently, there is no clear system for classifying different biorefinery concepts. However, they are often categorized based on feedstock, technology type, and technology status (van Ree & Annevelink, 2007). The depletion of natural resources and fossil fuels, coupled with the increasing demand for energy, has driven the development of sustainable and economically viable waste bioconversion processes, particularly for converting food waste into valuable bioproducts (Ahmed et al., 2023). The circular bioeconomy concept further promotes these processes, making sustainable bioproducts and bioconversion methods attractive to investors and producers. However, certain challenges, such as a lack of precise life cycle information for food biowaste and the need for robust techno-economic and environmental analyses, have hindered the widespread commercial adoption of this approach (Awasthi et al., 2021). The global demand for energy is on the rise due to factors like population growth, economic expansion, urbanization, and advancements in technology. This escalating energy consumption is expected to increase by 14% in OECD countries and 84% in non-OECD countries by 2035 (Wolfram et al., 2012). Consequently, the world's energy demand is increasing daily due to factors such as population growth, carbon emissions causing global warming, and increased waste generation (Kothari et al., 2010). In response, the waste-to-energy concept has been developed as an economically viable means to generate renewable energy from unusable waste products. This approach not only has economic feasibility but also offers environmental benefits by facilitating waste management to meet the world's energy demands.

Biorefineries are gaining importance as environmentally friendly alternatives to fossil fuels (Aragonés et al., 2022). The escalating demand for energy and resources is driving the transition from a fossil fuel-based linear economy to a more sustainable circular bioeconomy. The circular economy concept revolves around recycling formerly discarded materials into new products (Mhatre et al., 2021). The growing awareness of the negative environmental impact of waste disposal has spurred the growth of the waste valorization sector. Waste biorefineries represent a means to shift from a linear economy to a circular bioeconomy. These industrial-scale facilities optimize and sustainably utilize biomass for the production of energy, fuels, materials, and chemicals (de Albuquerque et al., 2019). The key drivers behind biorefineries are their cost-effectiveness, contribution to climate change mitigation, economic development potential, and support for technological and rural development. The biorefinery concept is emerging as a pivotal component of the circular economy, offering a pathway to produce a diverse range of sustainable products from biomass (Lago et al., 2019). Various categories of biomass are available for these purposes, as illustrated in Fig. 6.2, describing the biorefinery concept. Following the process of biomass degradation and conversion, it can be transformed into fuel and highly valued outputs such as proteins and amino acids.

## 6.3 Microorganisms as Versatile Tools in Biorefineries

The diversity and prevalence of microorganisms are truly remarkable. Their remarkable ability to thrive in a wide range of ecosystems and their extraordinary metabolic capabilities make them the most adaptable and successful inhabitants on our planet. They can be found in various environments, including soil (up to 3 km deep), the ocean, freshwater bodies like lakes, rivers, and ponds, the air, rocks, deserts, polar ice, oil fields, underground coal mines, underwater hydrothermal vents, hot springs, acidic lakes, soda lakes, the Dead Sea, the human body, walls, rocks, animal bodies, compost piles, rumen, termite guts, corals, paddy fields, high-pressure environments, anaerobic settings, sea anemones, tube worms (hydrothermal vents), insect guts, root nodules (symbiotic bacteria), plant cells (endophytes), plant roots (mycorrhiza), insect cells (endosymbionts), computer keyboards, subways/tunnels, and smartphones (Brock, 1967; Ley et al., 2006; Sirohi et al., 2012; Bunge et al., 2014; Clarke, 2014; Canganella & Wiegel, 2014; Mazard et al., 2016; Martin & McMinn, 2018). There are even reports of several species of bacteria and fungi surviving inside the modules of space stations.

The microbial diversity found in these diverse ecosystems serves as a valuable resource of microorganisms with industrial potential. Bioprospecting aims to identify such microorganisms and establish their economic and industrial significance. Advances in microbial ecology and biotechnology offer innovative and alternative approaches to traditional linear economic models. Bioprospecting and microbial-based solutions have the potential to support sustainable industries and contribute to global sustainable development goals. Notable examples include:

(i) Identifying new biosynthetic pathways in microbial genomes for novel drug production (Hemmerling & Piel, 2022).
(ii) Utilizing microorganisms to convert agro-industrial waste into valuable products, such as biofuels, biochemicals, and bio-based materials, reducing reliance on fossil fuels (Gonçalves et al., 2020).
(iii) Exploring microalgae capable of sequestering CO2 for bioproduct production and potentially serving as a source of landless food production, thus reducing the need for extensive agricultural land (Ullmann & Grimm, 2021). These microbial-based solutions contribute to improving human well-being and addressing long-standing environmental challenges that have contributed to global biodiversity loss.

In different aerobic and anaerobic communities, containing various fungal and bacterial species, the degradation of lignin-derived biomass has been observed (Tuomela et al., 2000). Mixed bacterial cultures have also demonstrated lignin degradation, and metagenomic studies have identified auxiliary activity families and metabolic pathways related to aromatic compounds (Jiménez et al., 2016; Silva et al., 2021). It has been proposed that bacteria can break down low molecular weight compounds produced during fungal lignin depolymerization. However, it remains unclear to what extent bacteria can depolymerize larger compounds, including high molecular weight lignin. Some studies suggest that bacteria have significantly lower lignin-degradation capabilities compared to fungal systems (Brown & Chang, 2014). If the prevailing view is that both fungal and bacterial species are necessary for complete lignin mineralization, it should be possible to identify metabolites that can only be consumed by specific community members and leverage synergistic interactions to enhance the conversion process. However, drawing such conclusions from the scientific literature remains challenging due to variations in lignin sources, organisms, and cultivation conditions, which influence community composition and the extent of lignin depolymerization (Azubuike et al., 2022). Certain microbes have been shown to degrade specific C–C and C–O bonds in lignin while altering the composition of H, G, and S lignin moieties (Zhu et al., 2022).

Microbial systems have also been employed recently to produce biofuels (such as bioethanol, biobutanol, biodiesel), bulk chemicals (e.g., lactic acid, 2-keto l-gulonic acid), and natural products like biopolymers (Jiang et al., 2020). Microbial consortia interact through mechanisms like synergism, commensalism, competition, and mutualism. Understanding these interactions, including the exchange of intermediate metabolites and cell-to-cell communication, can guide the design of artificial microbial consortia and enhance the robustness and stability of co-cultivation systems.

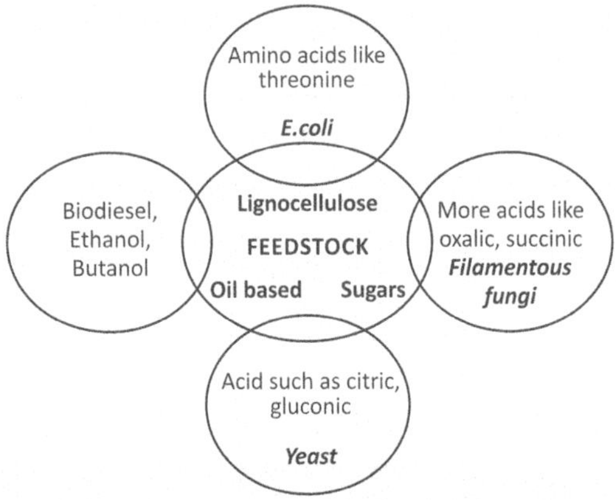


**Fig. 6.1** Example of biofuels and chemicals produced from renenwable feedstocks by microbial strains

Well-characterized microorganisms like *Saccharomyces cerevisiae* (yeast), *Escherichia coli* (bacteria), and other commercially employed microorganisms are often the first choice for developing novel biocatalysts for industrial applications. Figure 6.1 illustrates a selection of microbial strains used in biorefinery processes to produce valuable materials.

The yeast *S. cerevisiae* is widely recognized as the primary eukaryotic model microorganism, extensively employed in the bioethanol, brewery, and bakery industries across the globe. Moreover, ongoing research is paving the way for novel bioprocesses aimed at fuels and chemicals production, all centered around this yeast (Fischer et al., 2008). Its remarkable characteristics include a high ethanol tolerance, enabling minimal or no product inhibition, adaptability to diverse aeration conditions, including strict anaerobic environments, modest nutritional requirements, remarkable resilience to toxic compounds, and the ability to thrive in low-pH conditions, thereby mitigating the risk of bacterial contamination (Van Maris et al., 2007).

On the prokaryotic front, *E. coli* is the quintessential model organism. It shares similarities with yeast in terms of comprehensively understood genetics, physiology, and biology, alongside a well-established suite of genetic manipulation tools. *E. coli* has been genetically engineered for the production of a wide array of chemical compounds, showcasing its tremendous biotechnological potential (Buschke et al., 2013). With attributes highly coveted by the industry, including efficient growth under industrial conditions, minimal nutritional demands, anaerobic growth capabilities, and the versatility to utilize various carbon sources such as carbohydrates, polyols, and fatty acids (Clomburg & Gonzalez, 2010), *E. coli* has already been harnessed to produce ethanol from lignocellulose (Ingram et al., 1987). Currently, it is actively employed in a bioethanol pilot plant in Florida (Paes & Almeida, 2014).

## 6.4 Biofuel Production and Microbial Biocatalysts

The process of bioconversion, which transforms organic materials into energy sources or usable products, relies on biological agents and processes, particularly certain microorganisms (Maurya et al., 2021). One of its key advantages is the ability to convert biomass into pumpable slurry, thus reducing pre-processing costs and streamlining production. Several contemporary bioprocessing methods, such as microbiological and enzymatic approaches, bio-acidogenesis, electro-fermentation, anaerobic fermentation, oleaginous metabolism, and solventogenesis, have proven to be successful and popular choices for the sustainable and cost-effective conversion of biomass into value-added bioproducts (Torres-León et al., 2021). These bioproducts encompass a diverse range, including butanol, biogas, methane, hydrogen, ethanol, and biohythane (Atabani et al., 2022). Several of these are explored in further detail below:

### *6.4.1 Bioethanol*

This well-known renewable biofuel is an environmentally friendly alternative to fossil fuels. Bioethanol production primarily involves the hydrolysis of cellulose from lignocellulosic biomass and the fermentation of sugars derived from various lignocellulosic sources (Vasić et al., 2021). Aside from its biodegradability, bioethanol offers the advantage of lower toxicity, enabling the use of biomass as a primary substrate (Saha et al., 2017). Lignocellulosic materials, broadly categorized into marine algae, forest woody feedstocks, and agricultural residues, including municipal solid wastes, serve as the primary sources for bioethanol production. Figure 6.2 provides a schematic representation of the ethanol production process using lignocellulosic biomass.

Bioethanol production involves several stages, with the pretreatment process being the initial step. This method aims to alter the structural properties of the raw material to enhance enzyme accessibility and generate a significant amount of sugar monomers. Subsequently, enzymatic hydrolysis breaks down hemicellulose, cellulose, and starch. During this process, hexoses and pentoses are released and can be utilized in the later fermentation stage (Mirfakhar et al., 2020). Microorganisms digest the newly available sugars in the fermentation process, producing ethanol that is later distilled (Vasić et al., 2021). Enhanced fermentation techniques and specific microbial strains can maximize ethanol yields. Genetic modification of yeast and bacteria can also improve the efficiency of ethanol production from biomass. Recent research has indicated a peak ethanol yield of 7.9 g/L. The ideal conditions for achieving this yield include a bacterial biomass to fungal biomass ratio of 1:1, an initial pH of 5.5, and a solid loading of 5% (pretreated organic fraction of MSW) (Ebrahimian et al., 2022).

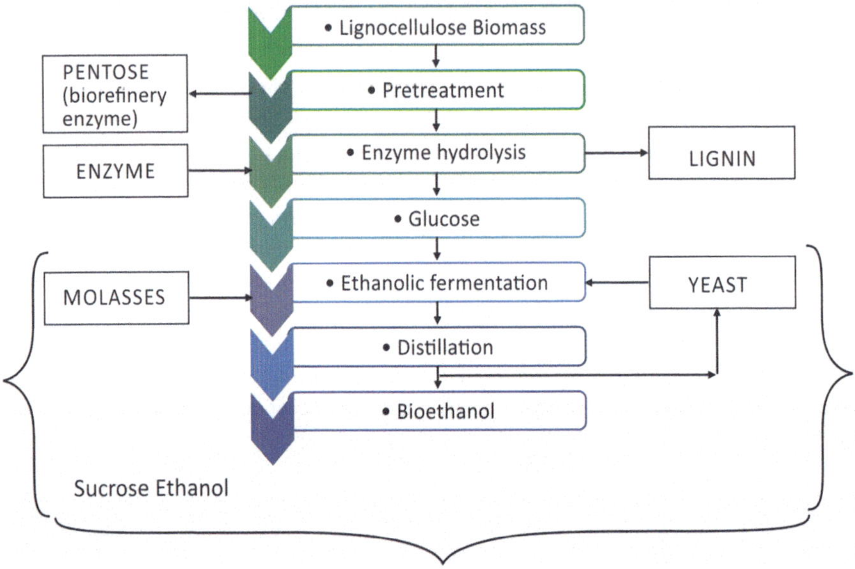


**Fig. 6.2** Process of bioethanol production from lignocelluloses biomass (Vasić et al., 2021)

### *6.4.2 Biobutanol as a Superior Biofuel*

Biobutanol is considered a superior biofuel compared to other alternatives such as bioethanol, biodiesel, and biohydrogen due to its chemical composition resembling gasoline (Ibrahim et al., 2018). It has garnered significant attention in recent years. Biobutanol can be produced from similar sources as bioethanol. It is relatively less polluting and more sustainable than gasoline. Recent findings by Li et al. (2020) emphasize that biobutanol exhibits superior vapor pressure behavior without an azeotrope, a higher energy density, remains in a single phase at low temperatures, has high viscosity, and a slightly lower octane number. Other notable advantages of biobutanol as a biofuel include its non-corrosive nature, lower volatility, increased energy input, and its compatibility with modern pipeline distribution without modification. Consequently, biobutanol has the potential to entirely replace the gasoline used for transportation (Ibrahim et al., 2018). There are two methods for biobutanol production: (i) the traditional ABE fermentation and (ii) microbial fermentation using industrial, residential, and agricultural wastes (Rathour et al., 2018). Table 6.1 provides an overview of the feasible substrates and fermentation conditions employed in various studies for biobutanol production.

**Table 6.1** Feasible substrate/feedstock and microbes utilized during the production of butanol

| Substrate feedstock | Fermentation condition | Microorganism | ABE (g/L) | BuOH (g/L) | Reference |
|---|---|---|---|---|---|
| Rotten potato | 96 h | *Clostridium beijerinckii DSMZ 6422* | 12.0 | 7.9 | Aves et al. (2023) |
| Green coconut husk | 96 h | *C. beijerinckii ATCC 10132* | 530 | 3.4 | De de brito bezerra et al. (2023) |
| Switchgrass | 72 | *C. beijerinckii SDR* | 20.24 | 11.21 | Olorunsogbon et al. (2022) |
| Cauliflower waste | Batch fermentation, 86 h | *C. acetobutyllum NRRL B527* | 5.29 | 2.99 | Khedkar et al. (2017) |
| Municipal solid waste | ABE fermentation | C. *acetobutyllum* | 16.1 | 5.9 | Ebrahimian et al. (2023) |

### 6.4.3 *Biohydrogen ($H_2$)*

Biohydrogen ($H_2$) is an environmentally friendly and renewable fuel that can be generated from organic waste materials, attracting global attention. Recent articles emphasize the need for further research on $H_2$ production from organic waste products (Osman et al., 2020). Utilizing waste materials from diverse biological sources to create sustainable biohydrogen offers a promising solution to contemporary environmental challenges and promotes a carbon-neutral economy (Osman et al., 2020). Among the various methods, dark fermentation is recognized for its efficiency in decomposing organic waste and achieving a high $H_2$ production rate (Osman et al., 2020; Yun et al., 2018). Combining anaerobic fermentation with photofermentation is an approach to enhance biological production by utilizing compounds produced during anaerobic fermentation. The synergy of dark fermentation and photofermentation has been shown to be the most effective means of achieving maximum $H_2$ production (Dahiya et al., 2018; Osman et al., 2020). Microalgae can also be a source of biohydrogen through fermentative and photobiological techniques, although pretreatment is necessary to optimize biohydrogen production (Ahmed et al., 2022). Anaerobic fermentation of microalgae presents an eco-friendly approach to generating renewable energy (Zhang et al., 2023). By leveraging the metabolic adaptability of specific microalgal species, these microorganisms can produce hydrogen gas even in oxygen-depleted conditions. Through cultivation in controlled environments and a pretreatment process, their hydrogen-producing potential becomes accessible. The biomass can be converted into hydrogen gas and valuable byproducts via anaerobic fermentation using specialized bioreactors (Sarangi & Nanda, 2020), offering a dual benefit of organic waste removal and clean fuel production.

### 6.4.4 Biohythane

The development of biohythane, a unique blend of H (5–25%) and $CH_4$ (75–95%) with enhanced fuel properties, is emerging as a promising and sustainable fuel option (Ghimire et al., 2022). Both $H_2$ and $CH_4$ come with their respective limitations; $H_2$ is reactive and poses storage challenges, while $CH_4$ has minimal flammability concerns (Dahiya et al., 2018). Biohythane, a combination of biohydrogen and biomethane, is produced at an industrial scale through a two-step fermentation process involving microalgae and biomass. A life-cycle assessment has been conducted to evaluate its environmental impacts and energy conversion properties (Sun et al., 2019). The system exhibits a net energy input/output ratio of 0.24. Approximately 53.8% of the total energy input is allocated to the biomass pretreatment process, while 16.6% is dedicated to microalgae cultivation.

### 6.4.5 Biodiesel

Biodiesel production can be sourced from a variety of lipid origins, including microalgae, oil crops, animal fats, and waste oils. Microalgae, in particular, have emerged as a sustainable feedstock for biodiesel due to their adaptability to thrive in diverse environments, such as saltwater, wastewater, and non-arable lands (Sani et al., 2013). This unique characteristic makes them an ideal choice, as their cultivation does not compete with food production or freshwater resources. Utilizing anaerobic fermentation techniques, microalgae can be cultivated and subjected to anaerobic conditions, leading to the fermentation process that produces biodiesel precursors like fatty acids and other valuable by-products (Zhang et al., 2022). This approach shows promise for environmentally friendly biodiesel production, capitalizing on microalgae's ability to convert light energy into biofuels. However, scaling this process to a commercial level necessitates further research and process optimization.

### 6.4.6 Biomethane

Biomethane, a refined type of biogas obtained by removing impurities (Chan Guti'errez et al., 2018), offers a comprehensive solution to numerous environmental and energy challenges, such as fossil fuel depletion, increased energy demands, inefficient waste management, and greenhouse gas (GHG) emissions, through biogas production via anaerobic digestion of biowaste (Batool et al., 2020). The utilization of biomethane in urban transportation has the potential to reduce greenhouse gas (GHG) emissions. By replacing diesel with biomethane, it is estimated that a 6.06% reduction in 2050 GHG emissions goals can be achieved (Batool et al.,

2020). Thanks to the efforts of various companies in designing and implementing advanced facilities, the commercial production of biomethane from food waste has become a sustainable and renewable biofuel on a global scale (Batool et al., 2020; Dahiya et al., 2018).

## 6.5 Enzymes Involved in Biofuel Synthesis and Their Sources

Biodiesel can be produced through three different methods: microemulsions, thermal cracking (pyrolysis), and trans/inter/esterification, as described by Ma & Hanna, 1999. Among these methods, the most cost-effective and high-quality approach is the transesterification of vegetable oils and animal fats. However, traditional acid/base transesterification processes come with several drawbacks, including high reaction temperatures, significant energy consumption, the corrosive nature of acids, the need for glycerol purification and recovery, the separation of catalysts and unreacted alcohol, and additional washing steps to remove impurities from biodiesel. These challenges can be mitigated or even eliminated by employing eco-friendly biocatalysts, such as enzymes, for biodiesel production.

Enzymes, particularly lipases, offer a biologically-based route to biodiesel production, offering numerous environmental and economic advantages compared to chemical methods, as discussed by Atadashi et al., 2011. These advantages include the use of lower reaction temperatures, high selectivity and specificity in trans/esterification processes, a broader range of substrates due to the ability to esterify both glyceride-linked and non-esterified fatty acids in a single step, reduced alcohol-to-oil ratios, avoidance of undesirable side reactions, simplified separation and product recovery through the production of a glycerol side stream with minimal impurities and water content (Du et al., 2008). The use of enzymes also eliminates the costs associated with recovering chemical catalysts, enhances enzyme biodegradability, and aligns with environmental sustainability.

## 6.6 Use of Lipase in Biodiesel Production

### *6.6.1 Enzyme Source*

Lipases, classified as hydrolases (acylglycerol acylhydrolases, EC 3.1.1.3), are enzymes that catalyze the hydrolysis of triglycerides (TG) into glycerol and fatty acids. They are versatile biocatalysts, capable of both hydrolysis and synthesis of long-chain acylglycerols, as defined by Jaeger et al., 1999. These lipases can be sourced from various origins and are categorized as plant, animal, or microbial lipases. Plant lipases have been identified in papaya latex, rapeseed, oat, and castor seeds (Sellappan & Akoh, 2005); however, they are not commonly used in commercial applications. In contrast, animal and microbial lipases are extensively utilized. Animal lipases are derived from the pancreases of cattle, sheep, hogs, and pigs (Sellappan & Akoh, 2005). Microbial lipases have gained significant industrial importance, currently accounting for approximately 5% of the world's enzyme market, following proteases and carbohydrases (Treichel et al., 2010). Microbial lipases are preferred due to their stability, availability in bulk, and lower cost compared to lipases from other sources.

Microbial lipases used as biocatalysts for biodiesel production are produced by various fungal, bacterial, and yeast species, as outlined in Table 6.2.

### 6.6.2 *Enzyme Characteristics*

Microbial lipases are predominantly extracellular enzymes with a molecular weight ranging from 30 to 50 kDa. They exhibit an optimal pH range in the slightly alkaline region, typically between 7.5 and 9 (Kakugawa et al., 2002). These lipases are primarily derived from mesophilic and thermophilic microorganisms, each with their distinct temperature preferences. Mesophilic lipases perform optimally within the temperature range of 35–50 °C, while thermophilic lipases exhibit their highest activity between 60–80 °C (Wang et al., 2007). It's noteworthy that mesophilic lipases tend to lose stability at temperatures exceeding 70 °C, whereas thermostable lipases can maintain their activity up to 100 °C in the presence of organic solvents and detergents (Antranikian, 2008). For example, lipases from hyperthermophilic archaea, such as *Pyrobaculumcalidifonti* (Hotta et al., 2002) and *Pyrococcusfuriosus* (Ikeda & Clark, 1998), as well as extreme thermophilic bacteria like *Thermoanaerobacterthermohydrosulfuricus* and *Caldanaerobactersubterraneus* (Royter et al., 2009), have been reported to display lipase activity at temperatures as high as 90 °C. Additionally, lipases from Thermoanaerobacter and Caldanaerobacter

**Table 6.2** Lipase-producing microorganisms

| Microbial type | Microbial source | Reference |
|---|---|---|
| Bacteria | *Acinetobacter calcoaceticus* | Pratuangdejkul and Dharmsthiti (2000) |
| | *Bacillus thermocatenulatus* | Vakhlu and Kour (2006) |
| | *Pseudomonas aeruginosa* | Ito et al. (2001) |
| | *Vibrio chloreae* | Sellappan and Akoh (2005) |
| Fungi | *Aspergillus fumigates* | Shangguan et al. (2012) |
| | *Penicillium cyclopium* | Chahinian et al. (2000) |
| | *Rhizopus oryzae* | Hiol et al. (2000) |
| | *Fusarium heterosporum* | Sellami et al. (2011) |
| Yeast | *Candida quercitrusa* | Shangguan et al. (2012) |
| | *Yarrowialipolytica NRRL YB-423* | Fickers et al. (2011) |
| | *Candida deformans* | Vakhlu and Kour (2006) |

have demonstrated resistance to organic solvents, making them promising candidates for biodiesel production in water-free environments.

Since lipases are mainly inducible enzymes, the primary determinant of their expression is the carbon source, which can either be a lipid source or another carbon-based compound (Lotti et al., 1998). Lipid sources encompass various substrates, including triacylglycerols, short or long-chain fatty acids, hydrolyzable esters, tween, bile salts, and glycerol (Gupta et al., 2004). Carbon sources may comprise sugars, sugar alcohols, polysaccharides, whey, amino acids, and other complex compounds. The incubation period required for maximum lipase production varies, typically ranging from a few hours to several days, depending on the specific organism and the chosen production method. Solid-state fermentation is the preferred method for cultivating fungal species, while bacteria and yeast are more efficiently produced through submerged fermentation (Dutra et al., 2008).

### *6.6.3 Biodiesel Conversion*

Lipases serve as catalysts for the trans/esterification reaction between triglycerides (TG) and acyl acceptors (alcohols). This reaction results in the formation of acyl-enzyme intermediates, which subsequently donate their acyl moiety to produce fatty acid alkyl esters (FAAE) (García-Alles & Gotor, 1998). The overall structure of lipases is characterized by a central β-sheet with the active serine located in a loop known as the catalytic elbow (Svendsen, 2000). Activation of lipase enzymes often requires the movement of a "lid." Some lipases, such as those in the *Thermomyceslanuginosus* family, possess an active site and a lid on the enzyme's surface. Others, like C. rugosa lipase, have their active site situated at the end of a tunnel containing the lid in its external parts (Pleiss et al., 1998). The structural variations among lipases from different sources can account for their differing activities on various oil substrates. Therefore, process optimization must be tailored to the selected enzymes and substrates for biodiesel production.

Furthermore, the choice of a specific lipase for lipid modification depends on the desired type of modification. It can be a position-specific modification of triacylglycerol, a fatty acid-specific modification, a modification by hydrolysis, or a modification by direct synthesis and transesterification (Bisen et al., 2010). Lipases can be categorized into three groups based on their substrate specificity: 1,3-specific, fatty acid-specific, and non-specific lipases. 1,3-specific lipases cleave ester bonds in positions 1 and 3 of TGs (Ribeiro et al., 2011). For example, *T. lanuginosus* 1,3-specific lipases have been reported to achieve a transesterification yield of 90% (Du et al., 2005). Fatty acid-specific lipases are known to hydrolyze esters of long-chain fatty acids with double bonds in the cis-position at C9, whereas non-specific lipases randomly cleave the acylglycerols into free fatty acids (FFA) (Ribeiro et al., 2011). For optimal biodiesel production, lipases should have the capacity to convert all three types of glycerides (mono-, di-, and triglycerides) into biodiesel, necessitating their non-stereospecificity and efficient catalysis of free fatty acid (FFA) esterification.

## 6.7 Case Studies Showcasing Microbial Biocatalysts in Biofuel Bioprocessing

Immobilization is the process of physically attaching enzymes to a solid support, enabling the conversion of substrates into products as they pass over the enzyme support. This technique is increasingly employed in industrial applications to streamline the separation of biocatalysts from the process stream, leading to improved product recovery and purification (Spahn & Minteer, 2008).

### *6.7.1 Enzyme Immobilization*

The immobilization of lipases has been achieved through various methods, including entrapment (Lee et al., 2009), physical adsorption (Rathi et al., 2001), ion exchange (Yang et al., 2006), and crosslinking (Lee et al., 2006). Various materials have served as carriers for lipase immobilization, such as polyurethane foam, silica, Sepabeads, and cellulosic nanofibers. The choice of immobilization technique and carrier depends on factors like the source of the lipase, the nature of the reaction system (aqueous, organic solvent, or two-phase), and the specific process conditions (pH, temperature, and pressure).

### *6.7.2 Whole Cell Immobilization*

In an attempt to avoid the complex enzyme recovery and purification requirements for immobilization of free (extracellular) lipase, immobilization of intracellular lipase, known as whole cell immobilization, has been extensively researched (Ban et al., 2001). Immobilization of the intracellular, cell- or membrane-bound lipase is believed to offer an alternative and less expensive source of immobilized biocatalyst for biodiesel production Lipase-producing bacteria, fungi and yeasts have been immobilized to serve as a lipase biocatalyst (Table 6.3). Matsumoto et al. (2001) developed whole cell biocatalysts by immobilizing *R. oryzae* cells, permeabilized by air drying. This biocatalyst was used in a three-step addition of methanol (known as methanolysis) to plant oil in a solvent- and water-free system at 37 °C. Some of the researcher reported that olive oil and oleic acid enhanced the methanolysis activity of immobilized *R. oryzae* cells with a step-wise addition of methanol (in presence of 15% water) yielding 90% biodiesel (Hama et al., 2007) Cross-linking of

**Table 6.3** Whole cell lipase biocatalysts used for biodiesel production

| Microbial source | Support material | Substrate | Conditions | Yield (%) | Reference |
|---|---|---|---|---|---|
| *P. Fluorescens* | Na-alginate | Jatropha oil | 40 °C; 3:1 methanol: Oil; hexane solvent; 48 h | 72 | Canakci (2007) |
| *A. Niger* | BSPs | Palm oil | 40 °C; 3:1 methanol: Oil; 3 step-wise; 72 h | 69 | Xiao et al. (2011) |
| *R. Orzyae* | BSPs + GA | Jatropha oil | 30 ° C; 3:1 methanol: Oil; 72 h | 89 | Tamalampudi et al. (2008) |
| *R. Mucilagenosa* | None | Palm oil | 30 °C; 3:1 methanol: Oil; 72 h | 51 | Srimhan et al. (2011) |
| *Pseudomonas sp.* | Na-alginate | Used cottonseed oil | 37 °C; 6:1 methanol: Oil; step-wise; 48 h | 70 | Ali et al. (2011) |

Na-alginate- sodium alginate. BSPs- biomass support particles. GA-glutaraldehyde

*R. oryzae* cells with 0.1% glutaraldehyde (GA) stabilized and maintained lipase activity at the same level even after six batches of methanolysis, with ester yields of 70–83% that represented a 20–33% increase over the control without cross-linking. Li et al. (2007a, b) studied biodiesel production from different feedstocks (refined, crude and acidified rapeseed oil) in tertiary-butanol using whole cell *R. oryzae* cells immobilized in biomass support particles (BSP)s. They found that the reaction rate and final ester yield were much greater when acidified rapeseed oil was used. Furthermore, the increase in the FFA in oil, the presence of phospholipids, and the use of adsorbents for water removal improved the rate of reaction and biodiesel yields. From this perspective, based on numerous literature reports, the lipasepro-ducing *R. oryzae*seems to show great promise for further research and development that could aid in reducing the cost of biodiesel production (Tamalampudi et al., 2008). A 72% biodiesel yield was obtained in transesterification of Jatropha oil with *P. fluorescens* MTCC 103 immobilized in sodium alginate gel. The optimum conditions were determined as follows: neutral pH, 40 °C, oil/ methanol mole ratio of 1:4, 48 h (Marchetti et al., 2008). *Pseudomonas sp.* was used as a whole cell biocatalyst for biodiesel production from used cottonseed oil: the FAME yield of 70% was attained by a step-wise addition of methanol in excess (oil/methanol mole ratio of 1:6) for 48 h (Ali et al., 2011). Pseudomonas sp. immobilized in sodium alginate gel was recommended for industrial use (Canakci, 2007). Lu et al. (2007) studied the transesterification of lard using immobilized Candida sp. 99–125 whole cells via a three-step methanolysis. They reported that for processing 1 g of lard the optimum, loading conditions are 0.2 g immobilized whole cell, 8 ml n-hexane as solvent, 20% water, and 40 °C. A biodiesel yield of 87.4% was obtained and the whole cell lipase was stable for 180 h of repeated use.

In addition to *R. oryzae,* other lipase-producing fungal and yeast whole cell catalysts have been also investigated in biodiesel production: a newly isolated strain of *A. niger* JN (Xiao et al., 2010); a methanoltolerant yeast *R. mucilagenosa*

(Srimhan et al., 2011); *Pseudomonas sp.* including *P. fluorescens* MTCC 103 (Foidl et al., 1996).

### 6.7.3 Biodiesel from Algae by Enzymatic Transesterification

Biodiesel production from algae through enzymatic transesterification has gained significant attention due to its high yield potential. The third generation of biofuels is increasingly exploring the use of microalgae as a primary feedstock for biofuel production, including biodiesel, bioethanol, and biogas, owing to the exceptional qualities of microalgae (Tran et al., 2013). Microalgae could be the potential and most economical source of biodiesel because of its high yielding feedstock. Microalgae have emerged as an economically promising source for biodiesel, thanks to their remarkable oil content. In fact, microalgae can contain over 80% oil by weight in their dry biomass, with typical oil levels ranging from 20% to 50% in various microalgae species (Borugadda & Goud, 2012; Christi, 2007). Biodiesel synthesis from microalgae biomass has been achieved through various cultivation methods, including photoautotrophic, heterotrophic, and mixotrophic growth systems, using either pure or mixed cultures. Photoautotrophic cultivation can be carried out in open or closed systems, while heterotrophic growth is typically sustained in fermenters (Daroch et al., 2013).

One critical factor for achieving efficient lipid production is the choice of the carbon source. For instance, *Chlorella protothecoides* can grow either photoautotrophically or heterotrophically, but the use of acetate or glucose as a carbon source tends to result in significantly higher biomass and lipid content. Table 6.4 illustrates the application of lipase enzymes in the biodiesel production process from algae. Chlorella species are among the most popular and versatile microalgae used in biodiesel production due to their impressive production yields. As an example, Xiong et al. (2008) employed lipase from Candida sp. 99–125 to catalyze the conversion of oil from *Chlorella protothecoides* into biodiesel, achieving an impressive conversion rate of 98% within just 12 hours.

**Table 6.4** Utilization of lipase enzyme from algae for biodiesel production

| Raw material/oil | Name of enzyme | Lipase origin | Immobilized method | Solvent used | Oil extraction method | Yield/ conversion |
|---|---|---|---|---|---|---|
| *Microalgae (Chlorella protothecoides)* | Lipase | Candida sp. 99–125 | Adsorption | Methanol | Soxhlet | 98.15% (Li et al., 2007a, b) |
| *Microalgae (Chlorella vulgaris ESP-31* | Lipase | Burkholderia sp.C20 | Covalent binding | Methanol | Sonication | 97.3% wt of oil (Tran et al., 2012) |

## 6.8 Economic Benefits and Potential of Microbial Bioprospecting for Biorefineries

Extremophiles, particularly thermophiles, play a pivotal role in the field of biorefinery by offering innovative metabolic pathways and robust enzymes capable of serving as biocatalysts under challenging industrial conditions. Thermophiles are organisms thriving in temperatures ranging from 41 to 122 degrees Celsius, with their optimal growth temperatures typically falling between 60 and 108 degrees Celsius. These microorganisms have garnered significant attention within the extremophile group due to their status as the primary source of industrially important thermostable enzymes (Singh et al., 2011). A thermostable enzyme or protein is defined by its ability to withstand high temperatures, as evidenced by a high transition temperature (Tm) or an extended half-life at elevated temperatures (Böhme et al., 2019). These unique properties of thermostable enzymes make them highly valuable for diverse industrial applications (Fig. 6.3). For instance, thermostable enzymes derived from thermophilic bacilli have found extensive use across various commercial applications owing to their resilient catalytic activity and their ability to endure the heat generated in numerous biotechnological and industrial processes (Margaryan et al., 2018). Thus, thermophilic enzymes can efficiently

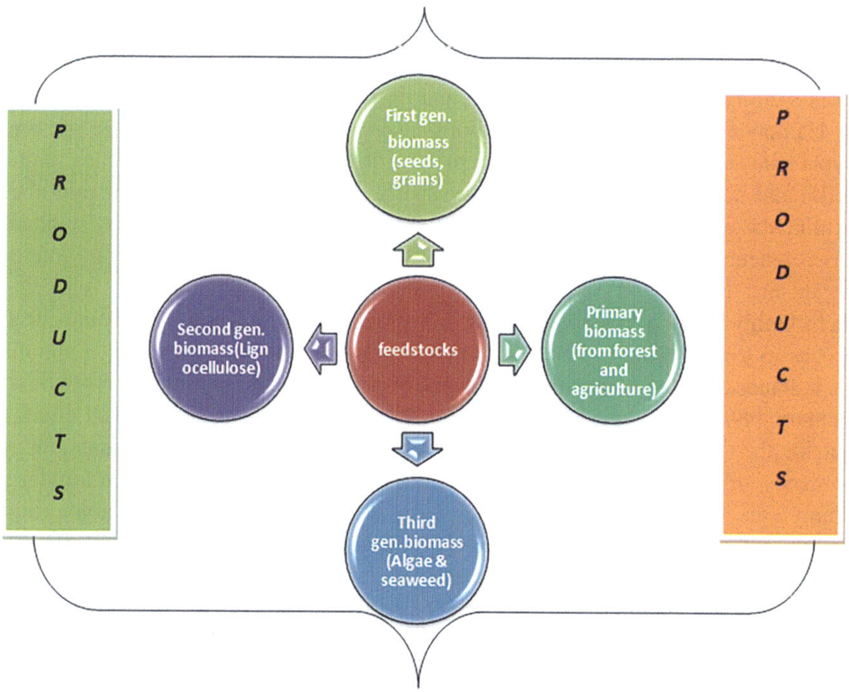


**Fig. 6.3** Various valuable products obtained from feedstock

catalyze high-temperature chemical processes that would challenge conventional enzymes designed for lower temperature conditions. One of the most renowned extremozymes is "Taq polymerase," originally isolated from the thermophilic bacterium "*Thermus aquaticus*" by F.C. Lowyer. Similarly, thermophilic Geobacillus species and Geobacillus sp. Iso5, isolated and characterized from thermal springs, have been found to produce hyperthermostable α-amylase with optimal enzyme activity at 90 and 140 degrees Celsius, respectively, making them exceptionally promising for biorefinery applications (Gurumurthy & Neelagund, 2012).

Moreover, various thermophilic bacteria and the enzymes they secrete have shown significant potential in bioenergy production and anaerobic fermentation. Examples include *Caldicellulosiruptorbescii, Geobacillusproteiniphilus, Thermoanaerobacterium, Pyrococcus, and Caldicellulosiruptor* (Hoffmann et al., 2020). Thermophilic Clostridium has successfully been employed as a sustainable green biotechnology solution for biobutanol production using corn stock as a substrate (Zhang & Jia, 2018). Notably, the introduction of Consortium TC-5 in digested sludge has been reported to substantially enhance methane production rates by 36.6% under thermophilic conditions, contributing to the sustainable development of the industrial economy (Kong, 2018).

Recent advancements have led to the expression of a glycocin biosynthetic gene cluster from the *thermophilic Aeribacillus pallidus* 8 in *E. coli*, offering potential applications in biofuel production (Kaunietis et al., 2019). Companies like Green Biologics have developed technologies to produce bio-butanol using corn raw materials and thermophilic Clostridium strains (Coker, 2016). Additionally, a xylanase produced by the acidophilic bacterium *Penicillium oxalicum* GZ-2 has demonstrated utility in biofuels, animal feed, and the food industries (Liao et al., 2012).

Halophiles are specialized microorganisms thriving optimally in saline and hypersaline environments, such as deep-sea sediments, saline lakes, salt pans, saline soils, and seawater (Kumar & Khare, 2012). In contrast, halotolerant microbes exhibit the ability to endure high-salt conditions while also thriving in standard environments. The inherent adaptability of these halophilic and halotolerant microorganisms depends on various environmental and nutritional factors, rendering them highly suitable for applications in biofuel production and various industrial processes (Singh et al., 2019). An example of a halophilic bacterium, Nesterenkonia sp. F, stands out for its capacity to withstand organic solvents and produce α-amylase. Remarkably, it's the sole wild bacterium known to produce butanol and ethanol outside the Clostridia genus. Further enhancing their industrial significance, many enzymes of halophilic and halotolerant bacterial origin, such as α-amylase, β-amylase, glucoamylase, lipase, esterase, hemicellulose, and ligninase, hold promise for applications in bioethanol, biodiesel, and fatty acids production using sustainable resources (Amoozegar et al., 2019).

In the realm of microalgae, diverse genera like Nannochloropsis, Dunaliella, Chaetoceros, Botryococcus, Scenedesmus, and Pseudochlorococcum have earned recognition for their capacity to accumulate substantial amounts of neutral lipids (Abomohra et al., 2016). The metabolic pathways of these algal strains can generate fatty acids with carbon chains ranging from 16 to 20 carbons, serving as precursors for biodiesel production. For instance, Nobre et al. (2013) harnessed Nannochloropsis

**Table 6.5** Products obtained from bioprospecting

| Species | Source | Enzyme | Application | Reference |
|---|---|---|---|---|
| *Anoxybacillus thermarum* | Hot spring | α-Amylase | Biorefinery, food, detergent | Mantiri et al. (2019) |
| Erwinia sp. E602 | Frozen soil | Cold-adapted β-galactosidase | Dairy industry | Xia et al. (2018) |
| *Exiguobacterium antarcticum* B7 | Antarctica soil | Cold-adapted β-galactosidase | Biorefinery, ethanol | Crespim et al. (2016) |
| Metagenome | Marine | Thermoactive endoglucanase | Biorefinery | Suleiman et al. (2019) |
| *Cellulomonas fimi* ATCC484 | Soil | Thermostable endoglucanase | Biorefinery | Saxena et al. (2018) |
| Thermophillic Anoxybacillus sp. BHH16 | Hot spring | Alkaline lipase | Wastewater treatment biodiesel | Burcu Bakir and Metin (2017) |
| *Paenicibacillus barengoltzii* | Marine | Chitinase | Conversion of cellulose to ethanol | Yang et al. (2016) |

*sp.* for fatty acid production used in biodiesel and other valuable co-products, such as carotenoids and bio-hydrogen. Their study showed that, through the use of CO2 supercritical fluid extraction, lipids constituting 45 g per 100 g of dry biomass and 70% of pigments could be effectively extracted.

In a more recent investigation, Kings et al. (2017) explored *Euglena sanguinea*, a microalga known for its high lipid production, as a candidate for biodiesel production. Analysis of the biodiesel revealed the presence of saturated fatty acids (C16:0, C18:0, C22:0, C24:0) and unsaturated fatty acids (C18:1), indicating their potential use in automobiles with minimal alterations to engine design. Table 6.5 shows the application of microbial bioprospecting.

Natural products (NPs) serve as vital reservoirs for novel bioactive compounds, including pharmaceuticals and high-value substances (Safonova et al., 2004). Traditionally, NPs have been sourced from plants and animals, yet their limited yield has impeded widespread utilization. Bacteria and fungi also hold significant promise as NP sources (Cragg & Newman, 2013). Notably, microbial consortia have been successfully employed in the production of NPs, exemplified by the cases of taxol and flavonoids, which exhibit considerable potential for pharmaceutical applications (Hooper et al., 2012). The biosynthetic pathway responsible for converting phenylpropanoic acids into flavan-3-ols comprises two distinct modules: an upstream module dependent on malonyl-CoA (converting phenylpropanoic acids to flavanones) and a downstream module reliant on NADPH (transforming flavanones into flavan-3-ols). However, when this intricate pathway was expressed within pure cultures, the yield of flavan-3-ols from phenylpropanoic acids remained notably low. In a groundbreaking effort, Chemler et al. (2007) engineered a binary system in *E. coli* that not only alleviated the metabolic burden but also allowed for the independent optimization of intermediate supply and co-factor provision within separate strains. Following comprehensive process optimization, which included factors like carbon source, temperature, induction points, and inoculation ratios, an impressive

40.7 mg/L of flavan-3-ols was achieved, marking a 970-fold improvement in flavonoid production compared to the pure culture approach (Jones & Kofas, 2016). For instance, the discovery of diorcinol J represented a significant milestone in NP exploration. This compound was synthesized from a marine isolate of the fungi *Aspergillus sulphureus* KMM 4640 and *Isariafelina* KMM 4639 (Zhuravleva et al., 2016).

## 6.9 Challenges and Future Directions

Technical challenges include covering feedstock yield, advanced technologies for lignin utilization, enzyme improvements, microbial cell factories, and processing and logistics.

(i) Enhancing the yield and composition of feedstock: It is imperative to increase both the yield and composition of biomass for optimal conversion efficiency. This necessitates advancements in plant genomics, breeding programs, and the engineering of desirable traits in biomass. By making feedstock more resilient, further enhancements can be made to the economic viability and availability of feedstock throughout the year.

(ii) Innovative approaches to utilize lignin: Lignin waste represents a significant byproduct in biofuel production, accounting for up to 30% of the weight and 40% of the energy value of biomass. Currently, this lignin is primarily incinerated to generate energy for biofuel plants. However, due to its chemical structure as a phenolic hetero-polymer, it can be transformed into higher-value products, such as carbon fiber, adhesives, phenolic resin, aromatics, and more, serving various industries. The advancement of technologies for producing these valuable products from lignin is essential for the success of the biorefinery industry.

(iii) Enhanced enzymatic efficiency: Another technical challenge is the development of more efficient and durable enzymes, particularly for converting lignocellulosic material derived from diverse feedstocks such as corncobs, stover, wheat straw, bagasse, rice, and woody biomass. Moreover, achieving greater utilization of the biomass requires innovative processes to extract their maximum value.

(iv) Microbial cell factories: An additional challenge related to yield improvement involves the creation of microbial cell factories capable of producing specific products with high yield and productivity. Novel heterogeneous catalysis technologies are also necessary to convert chemical intermediates into marketable products.

(v) Optimization of processing and logistics: The second set of technical challenges revolves around the optimization of feedstock processing and logistics. This encompasses various areas, including the development of densification techniques to facilitate cost-effective transport of initially low-density feedstock, the implementation of preservation methods to control physical and chemical changes in biomass during pre-conversion processing (e.g., harvesting, storage, transportation), and the establishment of a distribution network for bio-based products that leverages existing infrastructure, such as repurposing oil pipelines or upgrading fuel stations to accommodate a higher proportion of biofuels.

Therefore, a biodesign refinery should prioritize environmental sustainability, considering potential negative consequences such as competition with biomass and other raw materials, water usage, product quality, land utilization, carbon emissions, and impacts on biodiversity.

## 6.10 Conclusion

In the past two decades, many countries around the world have invested a lot of money into basic research and the technological development of first- and second-generation biofuels and bioenergy, respectively. In order to deal with the reduction of fossil energy reserves and the requirements of environmental protection, many biorefinery methods have been developed by researchers, but most of them stay in the laboratory or pilot stage; only a few have entered in industries for successful production. Though biofuel industries based on different technical routes have been built around the world, they still have to manage huge limitations in the form of production cost, maintenance, and technology development issues as well as an insufficient maturity of technology, high cost, and the lack of commercial competitiveness. At present, one of the biggest problems in biorefinery is that it has not yet achieved the full utilization of lignin, cellulose, and hemicellulose. One of the key bottlenecks in the green revolution is difficulties associated with biodegradation of lignin. These problems have to be addressed in a scientific, eco-friendly, and cost-effective way, which in turn will boost the industrial processes for sustainable production of biofuel. Extremozymes bring the dawn of success for biorefinery owing to their specificity, robustness in action, and high tolerance to the adverse conditions of the biorefinery process. It is expected that more extreme bacteria and their enzymes/proteins will provide more competitive chassis cells and catalysts for the production of biofuels in an efficient way. Hopefully, these current and future efforts will enable the conversion of various low-cost, abundant and environmental-friendly feedstock into advanced branched-chain fuels at high titers, yields, and productivities in industry-relevant scales.

## References

Abomohra, A. E., Jin, W., & El-Sheekh, M. (2016). Enhancement of lipid extraction for improved biodiesel recovery from the biodiesel promising microalga Scenedesmus obliquus. *Energy Conversion and Management, 108*, 23–29.

Ahmed, S. F., Mofijur, M., Nahrin, M., Chowdhury, S. N., Nuzhat, S., Alherek, M., et al. (2022). Biohydrogen production from wastewater-based microalgae: Progresses and challenges. *International Journal of Hydrogen Energy, 47*(88), 37321–37342.

Ahmed, S. F., Kabir, M., Mehjabin, A., Oishi, F. T. Z., Ahmed, S., Mannan, S., et al. (2023). Waste biorefinery to produce renewable energy: Bioconversion process and circular bioeconomy. *Energy Reports, 10*, 3073–3091.

Akondi, K. B., & Lakshmi, V. V. (2013). Emerging trends in genomic approaches for microbial bioprospecting. *OMICS: A Journal of Integrative Biology, 17*(2), 61–70.

Ali, A., Kaur, M., & Mehra, U. (2011). Use of immobilized pseudomonas sp. as whole cell catalyst for the transesterification of used cotton seed oil. *Journal of Oleo Science, 60*, 7–10.

Amoozegar, M. A., Safarpour, A., Akbari Noghabi, K., Bakhtiary, T., & Ventosa, A. (2019). Halophiles and their vast potential in biofuel production. *Frontiers in Microbiology, 10*, 1895. https://doi.org/10.3389/fmicb.2019.01895

Antranikian, G. (2008). Industrial relevance of thermophiles and their enzymes. In F. Robb, G. Antranikian, D. Grogan, & A. Driessen (Eds.), *Thermophiles: Biology and technology at high temperatures* (pp. 113–160). CRC Press.

Aragonés, M. M., Domínguez, C. G., Ondrejíčková, P., & Torralvo, F. A. (2022). Bioenergy production side-streams availability assessment as decision making driver for sustainable valorisation technologies development. Case study: Bioethanol and biodiesel industries. *Energy Reports, 8*, 6856–6865.

Atadashi, I., Aroua, M., & Aziz, A. A. (2011). Biodiesel separation and purification: A review. *Renewable Energy, 36*, 437–443.

Awasthi, M. K., Sarsaiya, S., Wainaina, S., Rajendran, K., Awasthi, S. K., Liu, T., et al. (2021). Techno-economics and life-cycle assessment of biological and thermochemical treatment of bio-waste. *Renewable and Sustainable Energy Reviews, 144*, 110837.

Azubuike, C. C., Allemann, M. N., & Michener, J. K. (2022). Microbial assimilation of lignin-derived aromatic compounds and conversion to value-added products. *Current Opinion in Microbiology, 65*, 64–72.

Ban, K., Kaieda, M., Matsumoto, T., Kondo, A., & Fukuda, H. (2001). Whole cell biocatalyst for biodiesel fuel production utilizing Rhizopus oryzae cells immobilized within biomass support particles. *Biochemical Engineering Journal, 8*, 39–43.

Barreto, L., Makihira, A., & Riahi, K. (2003). The hydrogen economy in the 21st century: A sustainable development scenario. *International Journal of Hydrogen Energy, 28*(3), 267–284.

Batool, N., Qazi, J. I., Aziz, N., Hussain, A., & Shah, S. Z. H. (2020). Bio-methane production potential assays of organic waste by anaerobic digestion and co-digestion. *Pakistan Journal of Zoology, 52*(3), 971.

Bisen, P. S., Sanodiya, B. S., Thakur, G. S., Baghel, R. K., & Prasad, G. (2010). Biodiesel production with special emphasis on lipase-catalyzed transesterification. *Biotechnology Letters, 32*, 1019–1030.

Bockris, J. O. M. (2003). On hydrogen futures: Toward a sustainable energy system. *International Journal of Hydrogen Energy, 1*(28), 131–133.

Böhme, B., Moritz, B., Wendler, J., Hertel, T., Ihling, C., Brandt, W., et al. (2019). Enzymatic activity and thermoresistance of improved microbial transglutaminase variants. *Amino Acids, 52*, 313–326. https://doi.org/10.1007/s00726-019-02764-9

Borugadda, V. B., & Goud, V. V. (2012). Biodiesel production from renewable feedstocks: Status and opportunities. *Renewable and Sustainable Energy Reviews, 16*(7), 4763–4784.

Brock, T. D. (1967). Life at high temperatures: Evolutionary, ecological, and biochemical significance of organisms living in hot springs is discussed. *Science, 158*(3804), 1012–1019.

Brown, M. E., & Chang, M. C. (2014). Exploring bacterial lignin degradation. *Current Opinion in Chemical Biology, 19*, 1–7.

Bunge, J., Willis, A., & Walsh, F. (2014). Estimating the number of species in microbial diversity studies. *Annual Review of Statistics and Its Application, 1*, 427–445.

Burcu Bakir, Z., and Metin, K. (2017). Production and characterization of an alkaline lipase from thermophilic Anoxybacillus sp. HBB16. Chem. Biochem. Eng. Q. 31, 303–312. https://doi.org/10.15255/CABEQ.2016.99

Buschke, N., Schäfer, R., Becker, J., & Wittmann, C. (2013). Metabolic engineering of industrial platform microorganisms for biorefinery applications–optimization of substrate spectrum and process robustness by rational and evolutive strategies. *Bioresource Technology, 135*, 544–554.

Canakci, M. (2007). The potential of restaurant waste lipids as biodiesel feedstocks. *Bioresource Technology, 98*, 183–190.

Canganella, F., & Wiegel, J. (2014). Anaerobic thermophiles. *Life, 4*(1), 77–104.

Chahinian, H., Vanot, G., Ibrik, A., Rugani, N., Sarda, L., & Comeau, L.-C. (2000). Production of extracellular lipases by Penicillium cyclopium purification and characterization of a partial acylglycerol lipase. *Bioscience, Biotechnology, and Biochemistry, 64*, 215–222.

Chemler, J. A., Lock, L. T., Kofas, M. A. G., & Tzanakakis, E. S. (2007). Standardized biosynthesis of favan-3-ols with efects on pancreatic beta-cell insulin secretion. *Applied Microbiology and Biotechnology, 77*, 797–807.

Christi, Y. (2007). Biodiesel from microalgae. *Biotechnology Advances, 25*(3), 294–306.

Clarke, A. (2014). The thermal limits to life on earth. *International Journal of Astrobiology, 13*(2), 141–154.

Clomburg, J. M., & Gonzalez, R. (2010). Biofuel production in Escherichia coli: The role of metabolic engineering and synthetic biology. *Applied Microbiology and Biotechnology, 86*, 419–434.

Coker, J. A. (2016). Extremophiles and biotechnology: Current uses and prospects. *F1000Research, 5*, F1000 Faculty Rev-1396. https://doi.org/10.12688/f1000research.7432.1

Cragg, G. M., & Newman, D. J. (2013). Natural products: A continuing source of novel drug leads. *Biochimica et Biophysica Acta, 1830*(6), 3670–3695.

Crespim, E., Zanphorlin, L. M., De Souza, F. H., Diogo, J. A., Gazolla, A. C., Machado, C. B., et al. (2016). A novel cold-adapted and glucose-tolerant GH1 β-glucosidase from Exiguobacterium antarcticum B7. *International Journal of Biological Macromolecules, 82*, 375–380. https://doi.org/10.1016/j.ijbiomac.2015.09.018

Dahiya, S., Kumar, A. N., Sravan, J. S., Chatterjee, S., Sarkar, O., & Mohan, S. V. (2018). Food waste biorefinery: Sustainable strategy for circular bioeconomy. *Bioresource Technology, 248*, 2–12.

Daroch, M., Geng, S., & Wang, G. (2013). Recent advances in liquid biofuel production from algal feedstocks. *Applied Energy, 102*, 1371–1381.

de Albuquerque, T. L., de Sousa Silva, J., de Macedo, A. C., Gonçalves, L. R. B., & Rocha, M. V. P. (2019). Biotechnological strategies for the lignin-based biorefinery valorization.

Dincer, I. (1999). Renewable energy and sustainable development: A crucial review. *Renewable and Sustainable Energy Reviews, 4*(2), 157–175.

Dincer, I., & Rosen, M. A. (2004). Exergy as a driver for achieving sustainability. *International Journal of Green Energy, 1*(1), 1–19.

Du, W., Xu, Y.-Y., Liu, D.-H., & Li, Z.-B. (2005). Study on acyl migration in immobilized lipozyme TL-catalyzed transesterification of soybean oil for biodiesel production. *Journal of Molecular Catalysis B: Enzymatic, 37*, 68–71.

Du, W., Li, W., Sun, T., Chen, X., & Liu, D. (2008). Perspectives for biotechnological production of biodiesel and impacts. *Applied Microbiology and Biotechnology, 79*, 331–337.

Dutra, J. C., Terzi, S. C., Bevilaqua, J. V., Damaso, M. C., Couri, S., Langone, M. A., et al. (2008). Lipase production in solid-state fermentation monitoring biomass growth of Aspergillus niger using digital image processing. *Applied Biochemistry and Biotechnology, 147*, 63–75.

Ebrahimian, F., Denayer, J. F., & Karimi, K. (2022). Efficient coproduction of butanol, ethanol, and biohydrogen from municipal solid waste through a cocultivated biorefinery. *Energy Conversion and Management, 255*, 115303.

EIA – u.s. (2017). *Energy Information Administration. Renewable energy explained*. https://www.eia.gov/energyexplained/index.cfm?Page=renewable_home. Accessed 27 June 2017.

Europabio. (2007). www.europabio.org

Fickers, P., Marty, A., & Nicaud, J. M. (2011). The lipases from *Yarrowialipolytica*: Genetics, production, regulation, biochemical characterization and biotechnological applications. *Biotechnology Advances, 29*, 632–644.

Fischer, C. R., Klein-Marcuschamer, D., & Stephanopoulos, G. (2008). Selection and optimization of microbial hosts for biofuels production. *Metabolic Engineering, 10*(6), 295–304.

Foidl, N., Foidl, G., Sanchez, M., Mittelbach, M., & Hackel, S. (1996). Jatropha curcas L. as a source for the production of biofuel in Nicaragua. *Bioresource Technology, 58*, 77–82.

García-Alles, L., & Gotor, V. (1998). Lipase-catalyzed transesterification in organic media: solvent effects on equilibrium and individual rate constants. *Biotechnology and Bioengineering, 59*, 684–694.

Ghimire, A., Luongo, V., Frunzo, L., Lens, P. N., Pirozzi, F., & Esposito, G. (2022). Biohythane production from food waste in a two-stage process: Assessing the energy recovery potential. *Environmental Technology, 43*(14), 2190–2196.

Gonçalves, C. C., Bruce, T., Silva, C. D. O. G., Fillho, E. X. F., Noronha, E. F., Carlquist, M., & Parachin, N. S. (2020). Bioprospecting microbial diversity for lignin valorization: Dry and wet screening methods. *Frontiers in Microbiology, 11*, 1081.

Gupta, R., Gupta, N., & Rathi, P. (2004). Bacterial lipases: an overview of production, purification and biochemical properties. *Applied Microbiology and Biotechnology, 64*, 763–781.

Gurumurthy, D., & Neelagund, S. (2012). Molecular characterization of industrially viable extreme thermostable novel alpha-amylase of geobacillussp Iso5 isolated from geothermal spring. *Pure and Applied Microbiology, 6*, 1759–1773.

Hama, S., Yamaji, H., Fukumizu, T., Numata, T., Tamalampudi, S., Kondo, A., et al. (2007). Biodiesel-fuel production in a packed-bed reactor using lipase-producing Rhizopus oryzae cells immobilized within biomass support particles. *Biochemical Engineering Journal, 34*, 273–278.

Hemmerling, F., & Piel, J. (2022). Strategies to access biosynthetic novelty in bacterial genomes for drug discovery. *Nature Reviews Drug Discovery, 21*(5), 359–378.

Hiol, A., Jonzo, M. D., Rugani, N., Druet, D., Sarda, L., & Comeau, L. C. (2000). Purification and characterization of an extracellular lipase from a thermophilic Rhizopus oryzae strain isolated from palm fruit. *Enzyme and Microbial Technology, 26*, 421–430.

Hoffmann, M., Gau, E., Braun, S., Pich, A., & Elling, L. (2020). Enzymatic synthesis of 2-(β-galactosyl)-ethyl methacrylate by β-galactosidase from Pyrococcuswoesei and application for glycopolymer synthesis and lectin studies. *Biomacromolecules, 21*, 974–987. https://doi.org/10.1021/acs.biomac.9b01647

Hooper, L., Kay, C., Abdelhamid, A., Kroon, P. A., Cohn, J. S., Rimm, E. B., & Cassidy, A. (2012). Efects of chocolate, cocoa, and favan-3-ols on cardiovascular health: A systematic review and meta-analysis of randomized trials. *The American Journal of Clinical Nutrition, 95*, 740–751.

Hotta, Y., Ezaki, S., Atomi, H., & Imanaka, T. (2002). Extremely stable and versatile carboxylesterase from a hyperthermophilic archaeon. *Applied and Environmental Microbiology, 68*, 3925–3931.

Ibrahim, M. F., Kim, S. W., & Abd-Aziz, S. (2018). Advanced bioprocessing strategies for biobutanol production from biomass. *Renewable and Sustainable Energy Reviews, 91*, 1192–1204.

Ikeda, M., & Clark, D. S. (1998). Molecular cloning of extremely thermostable esterase gene from hyperthermophilic archaeon Pyrococcusfuriosus in Escherichia coli. *Biotechnology and Bioengineering, 57*, 624–629.

Ingram, L. O., Conway, T., Clark, D. P., Sewell, G. W., & Preston, J. (1987). Genetic engineering of ethanol production in Escherichia coli. *Applied and Environmental Microbiology, 53*(10), 2420–2425.

Ito, T., Kikuta, H., Nagamori, E., Honda, H., Ogino, H., Ishikawa, H., et al. (2001). Lipase production in two-step fed-batch culture of organic solvent-tolerant Pseudomonas aeruginosa LST-03. *Journal of Bioscience and Bioengineering*, 91, 245–250.

Jaeger, K., Dijkstra, B. W., & Reetz, M. T. (1999). Bacterial biocatalysts: Molecular biology, three-dimensional structures, and biotechnological applications of lipases. *Annual Review of Microbiology, 53*, 315–351.

Jiang, Y., Dong, W., Xin, F., & Jiang, M. (2020). Designing synthetic microbial consortia for biofuel production. *Trends in Biotechnology, 38*(8), 828–831.

Jiménez, D. J., de Lima Brossi, M. J., Schückel, J., Kračun, S. K., Willats, W. G. T., & van Elsas, J. D. (2016). Characterization of three plant biomass-degrading microbial consortia by metagenomics- and metasecretomics-based approaches. *Applied Microbiology and Biotechnology, 100*, 10463–10477. https://doi.org/10.1007/s00253-016-7713-3

Jones, J. A., & Kofas, M. A. (2016). Optimizing metabolic pathways for the improved production of natural products. *Methods in Enzymology, 575*, 179–193.

Kakugawa, K., Shobayashi, M., Suzuki, O., & Miyakawa, T. (2002). Purification and characterization of a lipase from the glycolipid-producing yeast Kurtzmanomyces sp. I-11. *Bioscience, Biotechnology, and Biochemistry, 66*, 978–985.

Kaunietis, A., Buivydas, A., Čitavičius, D. J., & Kuipers, O. P. (2019). Heterologous biosynthesis and characterization of a glycocin from a thermophilic bacterium. *Nature Communications, 10*, 1115. https://doi.org/10.1038/s41467-019-09065-5

Kings, A. J., Raj, R. E., Miriam, L. R. M., & Visvanathan, M. A. (2017). Cultivation, extraction and optimization of biodiesel production from potential microalgae Euglena sanguinea using eco-friendly natural catalyst. *Energy Conversion and Management, 141*, 224–235.

Kong, X. (2018). Enhanced methane production from wheat straw with the assistance of lignocellulolytic microbial consortium TC-5. *Bioresource Technology, 263*, 7–39. https://doi.org/10.1016/j.biortech.2018.04.079

Kothari, R., Tyagi, V. V., & Pathak, A. (2010). Waste-to-energy: A way from renewable energy sources to sustainable development. *Renewable and Sustainable Energy Reviews, 14*(9), 3164–3170.

Kumar, S., & Khare, S. K. (2012). Purification and characterization of maltooligosaccharide-forming α-amylase from moderately halophilic Marinobacter sp. EMB8. *Bioresource Technology, 116*, 247–251. https://doi.org/10.1016/j.biortech.2011.11.109

Lago, C., Herrera, I., Caldés, N., & Lechón, Y. (2019). Nexus bioenergy–bioeconomy. In *The Role of Bioenergy in the Bioeconomy* (pp. 3–24). Academic Press.

Lee, D. H., Park, C. H., Yeo, J. M., & Kim, S. W. (2006). Lipase immobilization on silica gel using a cross-linking method. *Journal of Industrial and Engineering Chemistry, 12*, 777–782.

Lee, C.-H., Lin, T.-S., & Mou, C.-Y. (2009). Mesoporous materials for encapsulating enzymes. *Nano Today, 4*, 165–179.

Ley, R. E., Peterson, D. A., & Gordon, J. I. (2006). Ecological and evolutionary forces shaping microbial diversity in the human intestine. *Cell, 124*(4), 837–848.

Li, W., Du, W., & Liu, D. (2007a). Rhizopus oryzae IFO 4697 whole cell catalyzedmethanolysis of crude and acidified rapeseed oils for biodiesel production in tert-butanol system. *Process Biochemistry, 42*, 1481–1485.

Li, X., Xu, H., & Wu, Q. (2007b). Large-scale biodiesel production from microalga Chlorella protothecoides through heterotrophic cultivation in bioreactors. *Biotechnology and Bioengineering, 98*(4), 764–771.

Li, S., Huang, L., Ke, C., Pang, Z., & Liu, L. (2020). Pathway dissection, regulation, engineering and application: Lessons learned from biobutanol production by solventogenic clostridia. *Biotechnology for Biofuels, 13*(1), 1–25.

Liao, H., Xu, C., Tan, S., Wei, Z., Ling, N., Yu, G., et al. (2012). Production and characterization of acidophilic xylanolytic enzymes from Penicillium oxalicum GZ-2. *Bioresource Technology, 123*, 117–124. https://doi.org/10.1016/j.biortech.2012.07.051

Lotti, M., Monticelli, S., Luis Montesinos, J., Brocca, S., Valero, F., & Lafuente, J. (1998). Physiological control on the expression and secretion of Candida rugosa lipase. *Chemistry and Physics of Lipids, 93*, 143–148.

Lu, J., Nie, K., Xie, F., Wang, F., & Tan, T. (2007). Enzymatic synthesis of fatty acid methyl esters from lard with immobilized Candida sp. 99–125. *Process Biochemistry, 42*, 1367–1370.

Ma, F., & Hanna, M. A. (1999). Biodiesel production: a review. *Bioresource Technology, 70*, 1–15.

Mantiri, F., Rumende, R., & Sudewi, S. (2019). "Identification of α-amylase gene by PCR and activity of thermostable α- amylase from thermophilic Anoxybacillus thermarum isolated from Remboken hot spring in Minahasa, Indonesia," in IOP Conference Series: Earth and Environmental Science (Bristol: IOP Publishing). https://doi.org/10.1088/1755-1315/217/1/012045

Marchetti, J., Miguel, V., & Errazu, A. (2008). Techno-economic study of different alternatives for biodiesel production. *Fuel Processing Technology, 89*, 740–748.

Margaryan, A., Shahinyan, G., Hovhannisyan, P., Panosyan, H., Birkeland, N.-K., & Trchounian, A. (2018). Geobacillus and Anoxybacillusspp. from terrestrial geothermal springs worldwide: diversity and biotechnological applications. In D. Egamberdieva, N.-K. Birkeland, H. Panosyan, & W.-J. Li (Eds.), *Extremophiles in Eurasian ecosystems: Ecology, diversity, and applications* (pp. 119–166). Springer. https://doi.org/10.1007/978-981-13-0329-6_5

Martin, A., & McMinn, A. (2018). Sea ice, extremophiles and life on extra-terrestrial ocean worlds. *International Journal of Astrobiology, 17*(1), 1–16.

Maurya, D. K., Kumar, A., Chaurasiya, U., Hussain, T., & Singh, S. K. (2021). Modern era of microbial biotechnology: opportunities and future prospects. In *Microbiomes and plant health* (pp. 317–343). Academic Press.

Matsumoto, T., Takahashi, S., Kaieda, M., Ueda, M., Tanaka, A., Fukuda, H., et al. (2001). Yeast whole-cell biocatalyst constructed by intracellular overproduction of Rhizopus oryzae lipase is applicable to biodiesel fuel production. *Applied Microbiology and Biotechnology, 57*, 515–520.

Mazard, S., Penesyan, A., Ostrowski, M., Paulsen, I. T., & Egan, S. (2016). Tiny microbes with a big impact: The role of cyanobacteria and their metabolites in shaping our future. *Marine Drugs, 14*(5), 97.

Mhatre, P., Panchal, R., Singh, A., & Bibyan, S. (2021). A systematic literature review on the circular economy initiatives in the European Union. *Sustainable Production and Consumption, 26*, 187–202.

Midilli, A., Dincer, I., & Ay, M. (2006). Green energy strategies for sustainable development. *Energy Policy, 34*(18), 3623–3633.

Mirfakhar, M., Asadollahi, M. A., Amiri, H., & Karimi, K. (2020). Co-fermentation of hemicellulosic hydrolysates and starch from sweet sorghum by clostridium acetobutylicum: A synergistic effect for butanol production. *Industrial Crops and Products, 151*, 112459.

Nobre, B. P., Villalobos, F., Barragán, B. E., Oliveira, A. C., Batista, A. P., Marques, P. A. S. S., Mendes, R. L., Sovová, H., Palavra, A. F., & Gouveia, L. (2013). A biorefinery from Nannochloropsis sp. microalga – Extraction of oils and pigments. Production of biohydrogen from the leftover biomass. *Bioresource Technology, 135*, 128–136. https://doi.org/10.1016/j.biortech.2012.11.084

Osman, A. I., Deka, T. J., Baruah, D. C., & Rooney, D. W. (2020). Critical challenges in biohydrogen production processes from the organic feedstocks. *Biomass Conversion and Biorefinery*, 1–19.

Paes, B. G., & Almeida, J. R. (2014). Genetic improvement of microorganisms for applications in biorefineries. *Chemical and Biological Technologies in Agriculture, 1*(1), 1–10.

Pleiss, J., Fischer, M., & Schmid, R. D. (1998). Anatomy of lipase binding sites: The scissile fatty acid binding site. *Chemistry and Physics of Lipids, 93*, 67–80.

Pratuangdejkul, J., & Dharmsthiti, S. (2000). Purification and characterization of lipase from psychrophilic Acinetobacter calcoaceticus LP009. *Microbiological Research, 155*, 95–100.

Rathi, P., Saxena, R., & Gupta, R. (2001). A novel alkaline lipase from Burkholderiacepacia for detergent formulation. *Process Biochemistry, 37*, 187–192.

Rathour, R. K., Ahuja, V., Bhatia, R. K., & Bhatt, A. K. (2018). Biobutanol: New era of biofuels. *International Journal of Energy Research, 42*(15), 4532–4545.

Ribeiro, B. D., Castro, A. M., Coelho, M. A. Z., & Freire, D. M. G. (2011). Production and use of lipases in bioenergy: a review from the feedstocks to biodiesel production. *Enzyme Res*, Article ID 615803. https://doi.org/10.4061/2011/

Royter, M., Schmidt, M., Elend, C., Höbenreich, H., Schäfer, T., Bornscheuer, U., et al. (2009). Thermostable lipases from the extreme thermophilic anaerobic bacteria Thermoanaerobacterthermohydrosulfuricus SOL1 and Caldanaerobactersubterraneus subsp. tengcongensis. *Extremophiles, 13*, 769–783.

Safonova, E., Kvitko, K. V., Iankevitch, M. I., Surgko, L. F., Afti, I. A., & Reisser, W. (2004). Biotreatment of industrial wastewater by selected algal-bacterial consortia. *Engineering in Life Sciences, 4*(4), 347–353.

Saha, J., Begum, A., Mukherjee, A., & Kumar, S. (2017). A novel green synthesis of silver nanoparticles and their catalytic action in reduction of methylene blue dye. *Sustainable Environment Research, 27*(5), 245–250.

Sani, Y. M., Daud, W. M. A. W., & Aziz, A. A. (2013). Solid acid-catalyzed biodiesel production from microalgal oil—The dual advantage. *Journal of Environmental Chemical Engineering, 1*(3), 113–121.

Sarangi, P. K., & Nanda, S. (2020). Biohydrogen production through dark fermentation. *Chemical Engineering & Technology, 43*(4), 601–612.

Saxena, H., Hsu, B., De Asis, M., Zierke, M., Sim, L., Withers, S. G., et al. (2018). Characterization of a thermostable endoglucanase from Cellulomonas fimi ATCC484. Biochem. *Cell Biol. 96*, 68–76. https://doi.org/10.1139/bcb-2017-0150

Sellami, M., Aissa, I., Frikha, F., Gargouri, Y., & Miled, N. (2011). Immobilized Rhizopus oryzae lipase catalyzed synthesis of palm stearin and cetyl alcohol wax esters: Optimization by response surface methodology. *BMC Biotechnology, 11*, 68.

Sellappan, S., & Akoh, C. C. (2005). Applications of lipases in modifications of food lipids. In C. Hou (Ed.), *Handbook of industrial biocatalysis* (pp. 1–39). CRC Press LLC.

Shangguan, J.-J., Fan, L.-Q., Ju, X., Zhu, Q.-Q., Wang, F.-J., Zhao, J., et al. (2012). Expression and characterization of a novel enantioselective lipase from Aspergillus fumigatus. *Applied Biochemistry and Biotechnology, 168*, 1820–1833.

Silva, J. P., Ticona, A. R., Hamann, P. R., Quirino, B. F., & Noronha, E. F. (2021). Deconstruction of lignin: From enzymes to microorganisms. *Molecules, 26*(8), 2299.

Singh, S., Shukla, L., Nain, L., & Khare, S. (2011). Detection and characterization of new thermostable endoglucanase from aspergillus awamori strain F 18. *Journal of Mycology and Plant Pathology, 41*, 97–103.

Singh, P., Jain, K., Desai, C., Tiwari, O., & Madamwar, D. (2019). Microbial community dynamics of extremophiles/extreme environment. In *Microbial diversity in the genomic era* (pp. 323–332). Elsevier. https://doi.org/10.1016/B978-0-12-814849-5.00018-6

Sirohi, S. K., Singh, N., Dagar, S. S., & Puniya, A. K. (2012). Molecular tools for deciphering the microbial community structure and diversity in rumen ecosystem. *Applied Microbiology and Biotechnology, 95*, 1135–1154.

Spahn, C., & Minteer, S. D. (2008). Enzyme immobilization in biotechnology. *Recent Patents on Engineering, 2*, 195–200.

Srimhan, P., Kongnum, K., Taweerodjanakarn, S., & Hongpattarakere, T. (2011). Selection of lipase producing yeasts for methanol-tolerant biocatalyst as whole cell application for palm-oil transesterification. *Enzyme and Microbial Technology, 48*, 293–298.

Sun, C., Xia, A., Liao, Q., Fu, Q., Huang, Y., & Zhu, X. (2019). Life-cycle assessment of biohythane production via two-stage anaerobic fermentation from microalgae and food waste. *Renewable and Sustainable Energy Reviews, 112*, 395–410.

Suleiman, M., Schröder, C., Klippel, B., Schäfers, C., Krüger, A., and Antranikian, G. (2019). Extremely thermoactive archaeal endoglucanase from a shallow marine hydrothermal vent from Vulcano Island. *Applied Microbiology and Biotechnology, 103*, 1267–1274. https://doi.org/10.1007/s00253-018-9542-z

Svendsen, A. (2000). Lipase protein engineering. *Biochimica et Biophysica Acta, 1543*, 223–238.

Tamalampudi, S., Talukder, M. R., Hama, S., Numata, T., Kondo, A., & Fukuda, H. (2008). Enzymatic production of biodiesel from Jatropha oil: A comparative study of immobilized-whole cell and commercial lipases as a biocatalyst. *Biochemical Engineering Journal, 39*, 185–189.

Torres-León, C., Chávez-González, M. L., Hernández-Almanza, A., Martínez-Medina, G. A., Ramírez-Guzmán, N., Londoño-Hernández, L., & Aguilar, C. N. (2021). Recent advances on the microbiological and enzymatic processing for conversion of food wastes to valuable bioproducts. *Current Opinion in Food Science, 38*, 40-45.

Tran, D.-T., Yeh, K.-L., Chen, C.-L., & Chang, J.-S. (2012). Enzymatic transesterification of microalgal oil from Chlorella vulgaris ESP-31 for biodiesel synthesis using immobilized Burkholderia lipase. *Bioresource Technology, 108*, 119–127.

Tran, D.-T., Chen, C.-L., & Chang, J.-S. (2013). Effect of solvents and oil content on direct transesterification of wet oil-bearing microalgal biomass of Chlorella vulgaris ESP-31 for biodiesel synthesis using immobilized lipase as the biocatalyst. *Bioresource Technology, 135*, 213–221.

Treichel, H., de Oliveira, D., Mazutti, M. A., Di Luccio, M., & Oliveira, J. V. (2010). A review on microbial lipases production. *Food and Bioprocess Technology, 3*, 182–196.

Tuomela, M., Vikman, M., Hatakka, A., & Itävaara, M. (2000). Biodegradation of lignin in a compost environment: A review. *Bioresource Technology, 72*(2), 169–183.

Ullmann, J., & Grimm, D. (2021). Algae and their potential for a future bioeconomy, landless food production, and the socio-economic impact of an algae industry. *Organic Agriculture, 11*(2), 261–267.

Vakhlu, J., & Kour, A. (2006). Yeast lipases: Enzyme purification, biochemical properties and gene cloning. *Electronic Journal of Biotechnology, 9*, 68–85.

Van Maris, A. J., Winkler, A. A., Kuyper, M., De Laat, W. T., Van Dijken, J. P., & Pronk, J. T. (2007). Development of efficient xylose fermentation in Saccharomyces cerevisiae: Xylose isomerase as a key component. *Biofuels*, 179–204.

van Ree, R., & Annevelink, E. (2007). *Status report biorefinery 2007* (Report/Agrotechnology & Food Sciences Group; No. 847). Agrotechnology & Food Sciences Group. https://edepot.wur.nl/42141

Vasić, K., Knez, Ž., & Leitgeb, M. (2021). Bioethanol production by enzymatic hydrolysis from different lignocellulosic sources. *Molecules, 26*(3), 753.

Wang, L., Chi, Z., Wang, X., Liu, Z., & Li, J. (2007). Diversity of lipase-producing yeasts from marine environments and oil hydrolysis by their crude enzymes. *Annals of Microbiology, 57*, 495–501.

Wolfram, C., Shelef, O., & Gertler, P. (2012). How will energy demand develop in the developing world? *Journal of Economic Perspectives, 26*(1), 119–138.

World Economic Forum. (2010). *The future of industrial bio-refineries*. World Economic Forum.

Xia, Y., He, L., Mao, J., Fang, P., Ma, X., & Wang, Z. (2018). Purification, characterization, and gene cloning of a new coldadapted β-galactosidase from Erwinia sp. E602 isolated in northeast China. *Journal of Dairy Science, 101*, 6946–6954. https://doi.org/10.3168/jds.2018-14605

Xiao, M., Mathew, S., & Obbard, J. P. (2010). A newly isolated fungal strain used as whole cell biocatalyst for biodiesel production from palm oil. *GCB Bioenergy, 2*, 45–51.

Xiao, M., Qi, C., & Obbard, J. P. (2011). Biodiesel production using aspergillus Niger as a whole cell biocatalyst in a packed bed reactor. *GCB Bioenergy, 3*, 293–298.

Xiong, W., Li, X., Xiang, J., & Wu, Q. (2008). High-density fermentation of microalga Chlorella protothecoides in bioreactor for microbio-diesel production. *Applied Microbiology and Biotechnology, 78*(1), 29–36.

Yang, G., Tian-Wei, T., Kai-Li, N., & Fang, W. (2006). Immobilization of lipase on macroporous resin and its application in synthesis of biodiesel in low aqueous media. *Chinese Journal of Biotechnology, 22*, 114–118.

Yang, S., Fu, X., Yan, Q., Guo, Y., Liu, Z., and Jiang, Z. (2016). Cloning, expression, purification and application of a novel chitinase from a thermophilic marine bacterium Paenibacillus barengoltzii. *Food Chem. 192*, 1041–1048. https://doi.org/10.1016/j.foodchem.2015.07.092

Yun, Y. M., Lee, M. K., Im, S. W., Marone, A., Trably, E., Shin, S. R., Kim, M. G., Cho, S. K., Kim, D. H. (2018). Biohydrogen production from food waste: current status, limitations, and future perspectives. *Bioresource Technology*, https://doi.org/10.1016/j.biortech.2017.06.107

Zhang, J., & Jia, B. (2018). Enhanced butanol production using Clostridium beijerinckii SE-2 from the waste of corn processing. *Biomass and Bioenergy, 115*, 260–266. https://doi.org/10.1016/j.biombioe.2018.05.012

Zhang, S., Zhang, L., Xu, G., Li, F., & Li, X. (2022). A review on biodiesel production from microalgae: Influencing parameters and recent advanced technologies. *Frontiers in Microbiology, 13*, 970028.

Zhang, J., Xue, D., Wang, C., Fang, D., Cao, L., & Gong, C. (2023). *Genetic engineering for biohydrogen production from microalgae* (Vol. 26, p. 107255). Iscience.

Zhu, D., Qaria, M. A., Zhu, B., Sun, J., & Yang, B. (2022). Extremophiles and extremozymes in lignin bioprocessing. *Renewable and Sustainable Energy Reviews, 157*, 112069.

Zhuravleva, O. I., Kirichuk, N. N., Denisenko, V. A., Dmitrenok, P. S., Yurchenko, E. A., Min'Ko, E. M., Ivanets, E. V., & ShAfyatullov, S. (2016). New diorcinol J produced by co-cultivation of marine fungi aspergillus sulphureus and Isaria feline. *Chemistry of Natural Compounds, 52*, 227–230.

# Chapter 7
# Microbial-Based Recovery of Metals from E-Waste

**Contents**



## 7.1 Introduction

Heavy metals and other harmful compounds were released due to the widespread usage of electronic items, which is a severe environmental concern. The growing problem of "electronic waste," or "e-waste," which is made up of outdated electronic gadgets, is a result of consumer-oriented expansion, rapid product obsolescence, and technological advancements (Monika and Kishore 2010). Electronic trash management and safe disposal have grown in importance globally. For instance, the United Nations Environment Programme (UNEP) estimates that between 20 and 50 million metric tonnes of electronic garbage are produced and discarded annually worldwide (Narayanasamy et al., 2018). The United Nations (Forti et al., 2020) e-waste monitoring report estimates that the United States produces roughly 46 pounds of electronic garbage per person per year (Forti et al. 2020). The analysis calculates that 53.6 million metric tonnes of e-waste are created annually globally. Only around 17% of this waste is correctly collected, documented, and recycled worldwide each year, which may not come as a surprise but is nevertheless troubling. A large portion of the remaining e-waste, or 83%, is either



P. Saini, P. Mishra, *Bioprospecting of Microbial Resources for Agriculture, Environment and Bio-chemical Industry*, https://doi.org/10.1007/978-3-031-63844-2_7

disposed of or left sitting in households and workplaces. The central processing unit of electronic gadgets, printed circuit boards (PCBs), are made of a complex blend of metals, polymers, and ceramics. PCBs are thought to be a secondary source for several metals; merely a tiny portion of this is being appropriately recycled or handled.

A wide range of toxic organic and inorganic contaminants, including dioxins, heavy metals, polybrominated diphenyl ethers, polychlorinated biphenyls, brominated flame retardants, and more, are found in e-waste and can pose a threat to the whole environment including ecosystems, flora, and fauna. Heavy metals, acids, and deadly compounds are among the forms in which these toxins are either directly or indirectly discharged. Inhaling both fine and coarse particles, coming into contact with hazardous materials on the skin, or consuming contaminated dust can all result in direct exposure and consequent chances of hazard occurrence (Li & Achal, 2020). As e-waste moves up the food chain and comes in contact with humans through the food chain or other ways, it causes effects that become more concentrated and can result in a variety of health problems, including skin rashes, cancer, heart failure, birth deformities, mutationand DNA damage. (Li & Achal, 2020; Rautela et al., 2021). Numerous hazardous substances found in E waste are known to be harmful to human health or are suspected to do so. A number of these substances, including as dioxins, lead, and mercury, are among the 10 compounds of public health concern. According to estimates from the World Health Organization (WHO) and International Labour Organization (ILO), millions of child labourers and women employed globally in the informal recycling sector may be exposed to e-waste, and they may be especially sensitive given their developmental stage. (ILO, 2021). The World Health Organisation (WHO) published its first worldwide report on e-waste and child health in 2021, calling for more robust, effective measures to safeguard kids from the growing danger. E-waste recycling done poorly poses a risk to public health and safety.According to reports, the US e-waste recycling business formerly stated that around 90% of the e-waste it collected went to China and about 80% of it was exported throughout Asia (BAN, 2002). The recycling of electronic trash is a crucial topic from the perspectives of waste management and valuable material recovery. However, because of the complexity of the production equipment and the variety of materials found in the goods, recycling of electronic waste is still very limited. Currently, there is no accepted eco-friendly method for its secure disposal. Metals from e-waste have been mobilized in the past using a range of physicochemical techniques, such as hydrometallurgy and pyrometallurgy. However, the high operating costs brought on by the excessive use of energy and chemicals, significant metal loss, and secondary pollutant emissions have rendered these approaches ineffective. In developing nations, managing trash is a difficulty due to the recyclability of e-waste and the presence of potentially harmful pollutants. These wastes contain materials that are mined using antiquated technologies, which can harm both the environment and workers.

Global e-waste management is founded on the extended producer responsibility (EPR) concept. "Environmental product liability (EPR) is a strategy for protecting the environment in which the product's manufacturer assumes responsibility for the product's whole life cycle, particularly for its disposal, recycling, and take-back

(Lindquist, 2000)." The five factors; legal regulation, system coverage, system funding, producer accountability, and compliance that should be taken into account while developing an EPR-based e-waste management system were initially put forth by Lindhqvist. Using microbes to extract the metals from e-waste is an alternate strategy to address the issues raised above. For example, Cyanobacteria are a useful tool for removing silver and gold from PCBs. It has also been claimed that certain fungi can solubilize metals from PCBs by producing organic acids (Narayanasamy et al., 2018). A green and sustainable method, bio-hydrometallurgy (also called bioleaching) uses the metabolic processes of microorganisms (bacteria, fungi, and archaea) to solubilize metals from low-grade ores, mineral concentrates, or electronic trash. An alternative way to get over the issues mentioned above is to use microorganisms (bioleaching, biosorption, bioaccumulation, biotransformation, and biomineralization), which is a successful, affordable, sustainable process for solubilizing metals from e-waste (Sampath & Nigam, 2022). This method is mostly based on microbes' ability to produce lixiviants that mobilize metals. It is cost-effective, has few negative environmental effects, and produces little secondary waste (Dave et al., 2018). It can operate with diluted waste streams and at ambient temperatures and pressures. The physiology of the microorganisms and the mineralogical components of the e-waste are the primary factors that govern the leaching efficiency (Li et al., 2015). Consequently, the methods that microbes exploit to recover metals from e-waste are the main topic of this chapter. Furthermore, a thorough discussion is held on the bioprocess parameters that impact the effectiveness of metal bioleaching in conjunction with the methods employed for the profitable extraction of metals from electronic waste.

## 7.2 Sources of E-Waste

Electronic waste is one of the numerous new and significant waste sources that are adding to the continuously growing stream of solid waste. According to the Indian government, "e-waste" refers to electrical and electronic equipment that has been partially or completely thrown away as garbage by an individual or a large retailer, as well as rejected from the manufacturing, refurbishing, and repair processes (Kaur, 2018). E-waste originates from a wide range of sources, including individual users, homes, and organizations (Table 7.1). As a result, the two main categories of e-waste producers are individual (which encompasses households) and bulk consumers. As per the United Nations categorization outdated electrical and electronic devices are classified as electronic waste and fall into six categories: monitors (10%), televisions (10%), computers, phones, and related accessories (15%), DVD/VCR players, CD players, radios, and wi-fi sets (15%), refrigerators (20%), and laundry machines, air conditioners, vacuum cleaners, etc. (30%). A complex mixture of valuable metals (Cu, Ag, Au, and Pt) (Jiang et al., 2019); base metals (Cu, Al, Ni, Si, Zn, and Fe); hazardous metals (Hg, Be, Cd, Cr (VI), As, Sb, and Bi); and

**Table 7.1** Typical e-waste types and its sources

| S. N. | e-waste Types | Sources |
|---|---|---|
| 1. | Consumer electronics | This group comprises gadgets for use around the house or at home, including gaming consoles, audio equipment, tablets, computers, televisions, and cameras. |
| 2. | Technology and telecom equipment | Outdated or obsolete computer systems, servers, routers, switches, printers, scanners, fax machines, landlines, and mobile phones all produce e-waste. |
| 3. | Equipment used in office | This category covers outdated printers, fax machines, scanners, photocopiers, and other electronics used in offices. |
| 4. | Home appliances | Along with other electrical equipment in the home, appliances including air conditioners, microwaves, washing machines, dryers, and refrigerators also produce e-waste. |
| 5. | Electrical lighting equipment | The disposal of lightbulbs, fluorescent tubes, CFLs, and other lighting equipment adds to the amount of electronic waste produced. |
| 6. | Medical devices | When automated tool kits, various analyzers, X-ray machine, MRI, ultrasound,medical monitor, or other piece of technology used in a healthcare facility is no longer needed, it canbecome electronic trash. |
| 7. | Power equipment | Pertaining to includes transformers, batteries, solar panels, generators, uninterruptible power supplies (UPS), and other power-related gadgets. |
| 8. | Industrial equipment | A considerable quantity of e-waste can be produced by industrial machinery, such as automation systems, electronic control units, and equipment for manufacturing and processing. |
| 9. | Toysand recreational equipment | Electronic toys, video game consoles, electronic musical instruments, and other entertainment or recreational equipment can all be abandoned and produce e-waste. |
| 10. | Other electronic devices | A variety of other electronic products, including small electronic devices, electronic watches, calculators, and electronic cigarettes, are a source of e-waste. |

flammable materials (plastics, thermoplastics and flame retardants) make up the majority of electronic trash (Robinson, 2009).

## 7.3 Role of Microbes in Metals Recovery

A variety of biological processes were covered, including bioleaching, biosorption, bioaccumulation, biotransformation, and biomineralization, which include employing different microorganisms to recover metals. The development of methods for cultivating picky microorganisms, their identification, and the application of molecular biology techniques has made it evident that microorganisms from the three domains of life—bacteria, eukaryotes, and archaea—have potential roles and applications in biomining. The primary microbes implicated in the extraction of metals can be generically classified as oxidizers of iron and sulphur, cynogenic organisms, and makers of organic acids. Many microbes have different methods for bio-extraction of metals from solid waste sources (Narayanasamy et al., 2017, 2021). Some methods are discussed below:

- *Acidolysis*: Acidolysis is the process by which metals are liberated from a solid matrix by protonating the oxygen atoms in a metal oxide. By interacting with water molecules, oxygen atoms help to recover metals from their source and place them in a bioleaching media (Dutta et al., 2023).
- *Redoxolysis*: Redoxolysis is a technique that uses oxidation and reduction to change insoluble metals into their equivalent soluble metallic forms. Many chemolithotrophic bacteria, including, *Leptospirillum ferrooxidans*, *Acidithiobacillus ferrooxidans* and *Acidithiobacillus thiooxidans*, may involve in this process (Desmarais et al., 2020).
- *Complexolysis*: Another name for complexolysis is a chelation mechanism. When ligands and target metals from a solid matrix combine, cyanides, organic acids, or siderophores are formed. These compounds subsequently aid in the mobilization of metals from waste materials. After acidolysis, complexolysis takes place, which is how precious metals like platinum, gold, silver, iron, and palladium are extracted (Desmarais et al., 2020)
- *Biosorption*: Microorganisms from the bioleaching medium accumulate metals by a process known as biosorption, which is metabolically independent. It is a newly developed physico-chemical adsorption method that removes metals from electronic waste by complexation, chelation, coordination, and ion-exchange between the metals using specific kinds of dead, alive, or inactive microbial biomass. Various fungal, bacterial, and algal species' biomass has been employed as a biosorbent to extract heavy, precious, and rare earth metals from electronic trash (Ji et al., 2022).
- *Bioaccumulation*: In this microbe–metal interaction, soluble metals are mobilised intracellularly via the cell membrane, resulting in the accumulation of metal ions in the cell vacuole (Dusengemungu et al., 2021). The solubilization of metals from a solid matrix is facilitated by the presence of specific functional moieties; these include phosphate, amine, carboxyl, hydroxyl, etc., on the cell wall.

Oxidizers of iron and sulphur are important for the extraction of metals from minerals and e-waste. *Leptospirillum ferrooxidans, Acidithiobacillus thiooxidans*, and *Acidithiobacillus ferrooxidans* are the most well-known and thoroughly researched. *Chromobacterium violaceum, Pseudomonas fluorescens, Pseudomonas plecoglossicida*, and *Pseudomonas aeruginosa* are the main bacteria that have been linked to the generation of HCN. It has also been observed that fungi like *Polysphrous* and *Clitocyde* species of *Marsmius oreades* produce HCN. Given their capacity to alter metal speciation and boost metal mobility, these fungi may provide a viable substitute for current methods of removing metal from e-waste. Citric acid, oxalic acid, melic acid, and gluconic acid are examples of organic acids produced by fungi like *Aspergillus niger* and *Penicillium simplicissimum.*

## 7.4 Potential Benefits of Microbial Technology to Recover Metal from E-Waste

The primary benefits of microbial technology are its ability to operate in environments with diluted waste streams, at room temperature and pressure, at a low cost, with minimal secondary waste generation, and with reduced environmental impact. A viable and sustainable method for valuing e-waste and recovering valuable metals for use in a variety of industrial applications is called bioleaching. The methods outlined below can be used to extract metals from e-waste when microbes are present.

- *One step bioleaching*: A recently developed microbial culture is introduced into a sterile bioleaching medium in shaking flasks under aseptic conditions as part of a one-step bioleaching procedure (Wang et al., 2021).
- *Two step bioleaching*: This method involves cultivating microorganisms aseptically—without creating electronic waste—in shake flasks with new culture media under ideal bioprocess conditions until the organism reaches its maximum rate of cell growth and generation of biolixiviants. Next, as a second phase, sterilised e-waste at the appropriate pulp density and particle size is added to the culture medium, and it is further incubated for a specified amount of time (Dave et al., 2018).
- *Spent Medium Bioleaching*: In this method, an organism is grown in a culture medium in shake flasks in the absence of e-waste, until it attains peak cell density and biolixiviant production. Thereafter, e-waste is added to a cell-free culture supernatant obtained via centrifugation or filtration of the cultured broth and is subjected to further incubation (Liu et al., 2016)

Using microorganisms for bioleaching, metals from e-waste can be extracted effectively and with promise. The presence of suitable and ideal bioprocess conditions for the large-scale recovery of metals from e-waste determines the efficiency of the microorganisms.

## 7.5 Conclusion

The current e-waste treatment techniques continue to be hazardous and damaging to both humans and the surrounding biotic and abiotic environments. Diverse technologies are being developed and employed to extract metal from e-waste. Biohydrotechnology appears to be a safe and acceptable technique among these, although there are still gaps in the research that need to be filled. The two primary issues are that the organisms employed in the process were unable to survive for an extended period because of the toxicity of e-waste, and this approach is only capable of handling modest amounts of e-waste. To properly understand the distribution of communities, synergistic linkages, and mechanisms of action of a consortium or mixed strain of microbial organisms for the recovery of metals from e-waste, further research needs to be done. The use of non-traditional carbon sources, such as agro-industrial wastes like corncobs, rice bran, straw, peels, etc., ought to be promoted as an economical and environmentally beneficial medium for the growth of microorganisms for enhanced metabolite secretion and metal recovery from e-waste. Owing to the complexity of e-waste, which is made up of many different metals, further research should focus on using hybrid technologies to extract metals from waste materials more effectively and efficiently. To effectively recover metals from e-waste, a variety of leaching technologies such as biological, chemical, and physical processes must be integrated. Therefore, it is imperative to concentrate on developing new strains, expanding the current e-waste management and treatment process, optimizing existing strains, and contracting superior strains.

## References

BAN (The Basel Action Network), & Silicon Valley Toxicity (SVTC). (2002). *Exporting harms*. http://www.svtc.org/cleancc/pubs/technotrash

Dave, S. R., Sodha, A. B., & Tipre, D. R. (2018). Microbial technology for metal recovery from e-waste printed circuit boards. *Journal of Bacteriology and Mycology Open Access, 6*(4), 241–247.

Desmarais, M., Pirade, F., Zhang, J., & Rene, E. R. (2020). Biohydrometallurgical processes for the recovery of precious and base metals from waste electrical and electronic equipments: Current trends and perspectives. *Bioresource Technology Reports, 11*, 100526.

Dusengemungu, L., Kasali, G., Gwanama, C., & Mubemba, B. (2021). Overview of fungal bioleaching of metals. *Environmental Advances, 5*, 100083.

Dutta, D., Rautela, R., Gujjala, L. K. S., Kundu, D., Sharma, P., Tembhare, M., & Kumar, S. (2023). A review on recovery processes of metals from E-waste: A green perspective. *Science of the Total Environment, 859*, 160391.

Forti, V., Baldé C. P., Kuehr, R., & Bel, G. (2020). The Global E-waste Monitor 2020: Quantities, flows and the circular economy potential. United Nations University (UNU)/United Nations Institute for Training and Research (UNITAR) – co-hosted SCYCLE Programme, International Telecommunication Union (ITU) & International Solid Waste Association (ISWA), Bonn/Geneva/Rotterdam.

ILO. (2021). *ILO 2021 reports Child labour: Global estimates 2020, trends and the road forward*. International Labour Organization. https://www.ilo.org/ipec/ChildlabourstatisticsSIMPOC/lang%2D%2Den/index.htm

Ji, X., Yang, M., Wan, A., Yu, S., & Yao, Z. (2022). Bioleaching of typical electronic waste-printed circuit boards (WPCBs): A short review. *International Journal of Environmental Research and Public Health, 19*, 7508.

Jiang, B., Adebayo, A., Jia, J., Xing, Y., Deng, S., Guo, L., Liang, Y., & Zhang, D. (2019). Impacts of heavy metals and soil properties at a Nigerian e-waste site on soil microbial community. *Journal of Hazardous Materials, 362*, 187–195.

Kaur, H. (2018). Electronic waste management and its impact on Climate Change. In book: Socio Legal Dimensions of Climate Change. ENVIS Centre on Environmental Law & Policy, MoEFCC Publications.

Li, W., & Achal, V. (2020). Environmental and health impacts due to e-waste disposal in China—A review. *Science of the Total Environment, 737*, 139745.

Li, J., Liang, C., & Ma, C. (2015). Bioleaching of gold from waste printed circuit boards by *Chromobacterium violaceum*. *Journal of Material Cycles and Waste Management, 17*, 529–539.
Lindquist, T. (2000). *Extended producer responsibility in cleaner production*. The International Institute for Industrial Environmental Economics, Lund University.
Liu, R., Li, J., & Ge, Z. (2016). Review on *Chromobacterium violaceum* for gold bioleaching from e-waste. *Procedia Environmental Sciences, 31*, 947–953.
Monika & Kishore, J. (2010). E-waste management: as a challenge to public health in India. Indian journal of community medicine. *Indian Association of Preventive & Social Medicine, 35*(3), 382–385.
Narayanasamy, M., Dhanasekaran, D., & Thajuddin, N. (2017). Biological recovery of metals from electronic waste polluted environment usingmicroorganisms. In *Microbes for restoration of degraded ecosystems* (pp. 149–172). New India Publishing Agency.
Narayanasamy, M., Dhanasekaran, D., Vinothini, G., & Thajuddin, N. (2018). Extraction and recovery of precious metals from electronic waste printedcircuit boards by bioleaching acidophilic fungi. *International journal of Environmental Science and Technology, 15*(1), 119–132.
Narayanasamy, M., Dhanasekaran, D., & Thajuddin, N. (2021). Bioremediation of noxious metals from e-waste printed circuit boards by Frankia. *Microbiological Research, 245*, 126707.
Rautela, R., Arya, S., Vishwakarma, S., Lee, J., Kim, K. H., & Kumar, S. (2021). E-waste management and its effects on the environment and human health. *Science of the Total Environment, 773*, 145623.
Robinson, B. H. (2009). E-waste: an assessment of global production and environmental impacts. *Sci Total Environ, 408*(2), 183–191.
Sampath, M. K., & Nigam, V. K. (2022). Chapter 17: Microbial-based eco-friendly processes for the recovery of metals from E-waste. In P. Verma & M. P. Shah (Eds.), *Bioprospecting of microbial diversity* (pp. 393–405). Elsevier.
Wang, J., Faraji, F., Ramsay, J., & Ghahreman, A. (2021). A review of biocyanidation as a sustainable route for cold recovery from primary and secondary low-grade resources. *Journal of Cleaner Production, 296*, 126457.

# Chapter 8
# Assessing Technical and Commercial Aspects of Soil Microbiome and Microbial Enzymes in Formation of Bio-fertilizer and Environmental Sustainability

**Contents**



## 8.1 Introduction

To meet the world's food demand, intensive cropping and extensive use of water, agrochemicals, and mineral fertilizers resulted in environmental contamination, depletion of natural resources, and land degradation. Sustainable agriculture is based on producing food with as little of an impact on the environment and the food chain as possible from leftover chemical effects. The scientific community has long been interested in studying the normal relationship that plants have with their surroundings. Microbes and plants interact in a far more complicated way. Using soil microbiota to its full capacity in agriculture is thought to be the most effective long-term solution to problems with food security, environmental pollution, and human health (Suman et al., 2022). Anthropogenic activities, such as mining, agriculture, and waste management, are primarily responsible for soil contamination with heavy metals (HMs). The presence of HMs in inorganic fertilizers, pesticides, sewage sludge, and mining waste can disrupt soil microorganisms by seeping into the soil environment. It has been discovered that fast urbanization, along with reduced agricultural areas, sharp climate changes, and widespread use of agrochemicals in farming techniques, can lead to environmental disruptions and public health risks that compromise agricultural sustainability and food security. Because of their extensive history of application, the genus *Lactobacillus* and various *Lactobacillus* species

 
P. Saini, P. Mishra, *Bioprospecting of Microbial Resources for Agriculture, Environment and Bio-chemical Industry*, https://doi.org/10.1007/978-3-031-63844-2_8

are particularly valuable microorganisms (Bouis & Saltzman, 2017). Probiotic microorganisms are being utilised more frequently as dietary supplements in functional food products. Examples of these microbes include *Lactobacillus, Methanobrevibacter, Saccharomyces*, and *Bifidobacterium*. Microbes with advantageous qualities might be used for human health and sustainable agriculture. Another application for the microbiome is as a biocontrol agent. In addition to modulating phytohormone levels, the plant-associated microbiota (Kumar et al., 2018) produces siderophores, volatile chemicals, enzymes, and antibiotics, among other instances of biocontrol (Kumar et al., 2016; Tian et al., 2017). With their ability to promote plant growth, plant-associated bacteria hold great promise for resolving these issues and are essential for increasing agricultural production and biomass in both greenhouse and field settings. Less reliance on chemical pesticides and fertilizers is one of the direct benefits of the role of microbes in sustainable agriculture, as their ongoing use has not only shown negative effects on agro-ecosystems but also put human and animal health at risk. Because biofertilizers are less expensive, non-toxic, and environmentally benign, biofertilizers are an excellent alternative to costly and hazardous chemical fertilizers. This has given new insights into the agro-economy. In addition to harming human health and the ecological balance, a heavy reliance on chemical fertilizers to increase food production is becoming unaffordable for many farmers. One possible way out of this dilemma is to use helpful soil microorganisms in place of chemical fertilizers in food production. Bio-based fertilizers are essentially living cell preparations or latent potential microbial strains that, when applied to plants through soil or seed, help them absorb nutrients through their relationships in the rhizosphere. Rhizobacteria, or bacteria (e.g. *Rhizobium, Bradyrhizobium, Bacillus, Azospirillum, Azotobacter, Acinetobacter,* and *Pseudomonas*) that inhabit the rhizosphere, form unique paths of communication with plants and have the potential to affect plant physiology (Verma et al., 2017). Beneficial soil microorganisms improve rhizosphere management through a variety of multifaceted processes. These consist of siderophore production, nitrogen fixation, lytic acid generation, hydrogen cyanide production, phosphate solubilization, and indole acetic acid production. One of the biggest and most beneficial sources of several enzymes has been and still is microbes. Because of their proven ability to function as metabolic catalysts, microbial enzymes are used in a wide range of sectors and applications. The creation of commercial bioprocesses greatly depends on microbial enzymes. Pulp and paper, leather, detergents and textiles, medicines, chemicals, food and drink, biofuels, animal feed, and personal care are just a few of the various markets on which current applications are concentrated. The competitive global market for industrial enzymes is dominated by Novozymes, which is the largest player in the industry. DSM and DuPont are also strong contenders, having recently bought a controlling position in Danisco and its Genencor business (Adrio & Demain, 2014). An abundance of microbial enzyme resources is found in nature. Plant growth-promoting rhizobacteria (PGPRs), endo- and ectomycorrhizal fungi, cyanobacteria, and many other beneficial microscopic organisms can be applied in a variety of ways thanks to environmentally benign methods, which promote nutrient uptake, plant development, and plant tolerance to biotic and abiotic stress. The main obstacle in this field of study is that, in addition to identifying different microbial strains and their characteristics, it is crucial to analyze how these microbes and their enzymes perform to determine how effective they are at being used in sustainable agriculture. Numerous studies have demonstrated that by promoting plant growth and development, bacteria and arbuscular mycorrhizae fungus aid in the survival of plants under a variety of environmental stress conditions. In this chapter, an attempt was made to shed some light on the function of microorganisms, as well as how they affect crop yield and biotic/abiotic tolerance in sustainable agriculture. The goal of the chapter is to give readers an overview of the most recent uses of microbial enzymes in a variety of industrial areas.

## 8.2 Characteristic Features of Soil Microbiome

In the soil ecosystem, soil microbiota plays a unique and crucial role. It enhances plant health, quality, and fertility, all of which have a major impact on the yield and quality of agricultural products. According to Dubey et al. (2020), microbiota frequently comes into contact with complex organic, dead-decaying stuff that has significant energy and the nutritional value needed for cell maintenance and stable growth. Millions of microorganisms live in plants and soil, and these microorganisms come together to form a microbial community called the microbiome. A single soil can include a wide variety of microbial habitats, each with a unique microbial makeup. The pH, temperature, and availability of organic carbon in the soil are examples of soil and site factors that may often be used to predict the species abundances and composition of the bacterial communities in the soil (Prabhakaran et al., 2022). Sustainable agriculture depends on healthy soil quality, which is largely dependent on soil biology. In the rhizosphere, the small area around plant roots, the microbial makeup is more varied and plentiful. Typical rhizospheric soil contains up to $10^9$ cells per gram, representing up to $10^6$ species. In soil, fewer fungal and protist groupings are predominant, and most of the individual groups are little understood. According to Mahé et al. (2017), despite the ecosystem's diversity of species, parasitic members like Apicomplexa are the dominant species in Neotropical rainforests. The characteristics of the rhizospheric microbiome include promoting plant growth, suppressing illness, eliminating hazardous chemicals, and assisting plants in assimilating macro- and micronutrients. Using bio-formulations to enable such a favourable microbiome for crop production is an efficient method of regulating the overall crop output and plant quality (Prasad et al., 2021). The soil-rich communities inhabited by Bacteroidetes, Acidobacteria, Proteobacteria, Actinobacteria, and Planctomycetes are enhanced by the dynamic character of microbiome communities that are coupled with plant roots, ideally through horizontal transfer (Ecker et al., 2019). According to Jørgensen et al. (2019), sulphate-reducing bacteria (SRB) including six main categories-*Desulfotomaculum, Desulfobulbus, Desulfobacterium, Desulfobacter, Desulfococcus-Desulfonema-Desulfosarcina, and Desulfovibrio-Desulfomicrobium* are essential for the breakdown of organic complex.

## 8.3 Enzymes Derived from Microbes

Extracellular enzymes are a common term for microbial enzymes. Different organic material-degrading enzymes, including those that break down polysaccharides, pesticides, PAHs etc. are secreted by bacteria. Numerous applications, including bioremediation and biogeochemical cycling, benefit from biodegradation (Ortiz-Hernández et al. 2013). The following are a few of the most remarkable instances of microbial extracellular enzymes: proteases, ligninases, hyaluronidases, ligninases, cellulases, chitinases, hemicellulases, and so forth. Because microbial enzymes exhibit high and exceptional performance levels and can function effectively in a wide variety of diverse physical and chemical circumstances, their production is an essential process in the industrial sectors. Some of the microorganism-derived enzymes are given below in Table 8.1.

Enzymes are crucial for many processes of industrial, medicinal, and biotechnological importance. Natural enzymes have been utilized for millennia to produce cheese, wine, beer, and vinegar, among other food products (Kirk et al., 2002). Microbial enzymes are the ideal source for industrial applications because they are easier to genetically modify bacterial cells to produce more enzymes, and they can be produced in large quantities quickly and with shorter generation durations. More than 500 commercial goods are made using enzymes (Johannes & Zhao, 2006). Approximately 60% of the enzyme market worldwide is made up of proteases. They are employed in the manufacturing of agrochemicals, detergents, food and feed, leather, textile and silk goods. The largest market for industrial

**Table 8.1** Microorganism-derived enzymes

| Source | Enzyme | Microorganism |
|---|---|---|
| Fungi | Amylases | *Aspergillus oryzae* |
| | Glucosidases | *Aspergillus flavus* |
| | Proteases, Pectinases, Catalases, Nitrilase | *Aspergillus niger* |
| | Glucose oxidase | *Penicillium notatum* |
| | Peroxidase | *Phanerochaete, Chrysosporium, Phanerochaete laevis* |
| | Phosphatase | *Aspergillus* sp. |
| Bacteria | Nitroreductase | *Comamonas* sp., *Pseudomonas putida* |
| | Dehalogenase | *Xanthobacter, Autotrophicus, Sphingobium, Chlorophenolicum* |
| | Proteases, Penicillinase, Amylases | *Bacillus subtilis* |
| | Dioxygenase | *Pseudomonas* sp., *Mycobacterium* sp. |
| | laccase | *Streptomyces laccase, Pseudomonas putida* F6 |
| Yeast | Invertase | *Saccharomyces cerevisiae* |
| | Lactase | *Saccharomyces fragilis* |

enzymes is that of food and beverage enzymes, with sales of approximately $1.2 billion in 2011 and projected to reach $1.8 billion by 2016 at a calculated annual growth rate of 10.4% (Adrio & Demain, 2014). Currently in use in the food business are three recombinant fungal lipases generated in *A. oryzae*, one from *Rhizomucor miehi*, one from *Thermomyces lanuginosus*, and one from *Fusarium oxysporum* (Rao et al., 1998). Due to its high specificity, calf rennin was once preferred in the production of cheese; however, microbial proteases generated by Gram-positive lactic acid bacteria (GRAS) such as *Mucor miehei*, *Bacillus subtilis*, *Mucor pusillus Lindt*, and *Endothia parasitica* are rapidly displacing it. Currently, a $1 billion industry uses bacterial glucose isomerase, fungal α-amylase, and glucoamylase to convert starch into high fructose corn syrup. Other useful enzymes in the food industry are β-galactosidase (lactase) from *Kluyveromyces lactis*, *K. fragilis* or *Candida pseudotropicalis* for hydrolyzing lactose from milk or whey, and galactosidase from *S. carlsbergensis* for crystallizing beet sugar. Cellulase is produced by a vast and diverse fungus population, such as the genera *Trichoderma*, *Chaetomium*, *Penicillium*, *Aspergillus*, *Fusarium* and *Phoma*; aerobic bacteria, such as *Acidothermus*, *Bacillus*, *Celvibrio*, *pseudonoma*, *Staphylococcus*, *Streptomyces* and *anthomonas*; and anaerobic bacteria, such as *Acetovibrio*, *Bacteroides*, *Butyrivibrio*, *Caldocellum*,*Clostridium*, *Erwinia*, *Eubacterium*, *Pseudonocardia*, *Ruminococcus* and *Thermoanaerobacter* (Moreira & Siqueira, 2006). Invertase isolated from *Kluyveromyces fragilis*, *Saccharomyces carlsbergensis*, and *S. cerevisiae* utilized for manufacturing candy and jam. The fungi of the genera *Rhizopus*, *Geotrichum*, *Rhizomucor*, *Aspergillus*, *Candida* and *Penicillium* have been reported to be producers of several commercially used lipases.

MannawayTM is a *Bacillus mannanase* enzyme that was recently introduced to detergents to remove guar gum-containing food stains. The most recent commercial developments include the use of cellulases for denim finishing and laccases for bleaching and decolorizing textile effluents. The use of lipases, xylanases, and laccases in the pulp industry to remove pitch—a hydrophobic mixture of mostly triglycerides and waxes—is another increasingly significant usage of enzymes. Environmental contaminants and lignin-related chemicals, both phenolic and non-phenolic, are oxidised by laccases. It has been employed as a medical diagnostic tool, as a cleaning agent for water purification systems, as a pharmaceuticals catalyst, as a detoxifier for industrial effluents from the textile, paper and pulp,and petrochemical industries, and as a cosmetic ingredient. It has also been used for bioremediation of pesticides, herbicides, and explosives in soil. To convert lignocellulosic biomass into biofuels and other products at a reasonable cost, cellulases are essential. Phytases, proteases, α-galactosidases, glucanases, xylanases, α-amylases, and polygalacturonases are commercially available feed enzymes that are mostly utilized for swine and poultry feed (Selle & Ravindran, 2007).

## 8.4 Environmental Sustainability of Bio-fertilizer and Microbial Enzymes

Being a significant source of food, fibre, and biodiversity, soil is valued as a resource for agriculture, human health, and the environment. People are starting to realize more and more how their diet impacts their general health. But we often forget about the whole food life cycle, concentrating only on the latter stages of processing and neglecting the health of the soil, which is necessary to support the 95% of food that comes from the soil (Singh, 2019). A healthy society is derived from healthy soil. Because soil bacteria convert degradable organic substances into inorganic forms, they are essential in determining the nutritional value of our food.Application of beneficial bacteria successfully contributes to the protection of biodiversity as well as soil health, water holding capacity, carbon storage, root growth, availability and recycling of vital nutrients, and combat pollution.This method not only guarantees that crops will receive enough nutrients, but it also enriches the soil's organic matter and enhances a number of its physical characteristics. It is possible to separate, combine, and create microorganisms that play important roles in the microbiome into new biological products known as biofertilizers (Suman et al., 2022). Between 2020 and 2025, the Indian bio-fertilizer market is projected to expand at a compound annual growth rate (CAGR) of about 13%. Based on global biofertilizer usage, the biofertilizer industry is expected to reach a valuation of 2.3 billion US dollars by 2020. By 2025, it is projected to reach 3.9 billion USD at a CAGR of 11.6% (Kodgire et al., 2022). Mycorrhizal fungi are essential for improving the uptake of nutrients and water from the soil, particularly phosphorus, which is necessary for plant growth and productivity. Similar to this, it has been found that the inorganic phosphate transporter (Pi) in the hyphae of mycorrhiza *Glomus versiformis* improves the uptake of phosphate by the host plant from the soil.Pesticides and hydrocarbons have frequently been broken down by a variety of enzymes (such as Chironome P450, laccases, hydrolases, dehalogenases, dehydrogenases, proteases, and lipases) from aerobic bacteria, including *Pseudomonas*, *Alcaligenes*, *Sphingomonas*, *Rhodococcus*, and *Mycobacterium* (Bhandari et al., 2021). These are also responsible for the degradation of the many kinds of pollutants utilized in bioremediation procedures. A particular pollutant substrate is identified by a particular microbial enzyme under ideal conditions (temperature, pH, contact time, concentration), enabling it to effectively undergo enzymatic conversion into a variety of safe products through ring-opening, oxidation-reduction, and elimination. Understanding, the regulation of genes and the expression of extracellular enzymes is of tremendous interest due to the biotechnological significance of microbes and their enzymes.

## 8.5 Conclusion

Microorganisms are a remarkable source of catalysts with a broad range of uses in a variety of industries, including pharma, food, fine chemicals, household care, and animal feed. Because of the special qualities of enzymes like their high specificity, quick action, and biodegradability enzyme-assisted processes in industry can operate at softer reaction conditions, produce higher yields, and produce less waste. Because of their biodegradable nature, microorganisms and microbial enzymes are used in new environmentally friendly technologies that support sustainable development. These technologies can safely increase soil fertility without accumulating contaminants and preserve the integrity of the environment and its ecosystems. The use of biofertilizers is safe, affordable, and environmentally friendly. It also has a large potential to boost plant production. To meet the global population's demand for food production, the use of biofertilizers as a crucial part of agricultural practice to promote plant yield has gained increasing traction recently. Therefore, the role of microorganisms that promote plant growth and the use of biofertilizers derived from viable microbial strains in the field are favourable for the effective management of the rhizosphere in sustainable agriculture.

## References


Adrio, J. L., & Demain, A. L. (2014). Microbial enzymes: Tools for biotechnological processes. *Biomolecules, 4*(1), 117–139.

Bhandari, S., Poudel, D. K., Marahatha, R., Dawadi, S., Khadayat, K., Phuyal, S., Shrestha, S., Gaire, S., Basnet, K., Khadka, U., & Parajuli, N. (2021). Microbial enzymes used in bioremediation. *Journal of Chemistry, 2021*, Article ID 8849512, 17 pages.

Bouis, H. E., & Saltzman, A. (2017). Improving nutrition through biofortification: A review of evidence from HarvestPlus, 2003 through 2016. *Global Food Security, 12*, 49–58.

Dubey, R. K., Tripathi, V., Prabha, R., Chaurasia, R., Singh, D. P., Rao, C. S., Abhilash, P. C., et al. (2020). *Unravelling the soil microbiome: Perspectives for environmental sustainability*. Springer.

Ecker, G., Porto, L., Arias, R. S., Cano, L. M., Rott, P., & Urashima, A. S. (2019). *Sugarcane cultivars susceptibleto orange rust can host sub-populations of Puccinia kuehnii capable of infecting resistantcultivars*. The American Phytopathological Society.

Johannes, T. W., & Zhao, H. (2006). Directed evolution of enzymes and biosynthetic pathways. *Current Opinion in Microbiology, 9*(3), 261–267.

Jørgensen, B. B., Findlay, A. J., & Pellerin, A. (2019). The biogeochemical sulfur cycle of marine sediments. *Frontiers in Microbiology, 10*, 849.

Kirk, O., Borchert, T., & Fuglsang, C. (2002). Industrial enzyme applications. *Current Opinion in Biotechnology, 13*, 345–351.

Kodgire, S., Saxena, N., Chandrashekharaiah, P. S., Sanyal, D., & Dasgupta, S. (2022). Assessing technical and commercial aspects of soil microbiome in growing leguminous plants and formation of bio-fertilizer. https://doi.org/10.1016/B978-0-323-90958-7.00021-2

Kumar, P. A., Dubey, R. C., Maheshwari, D. K., & Bajpai, V. (2016). ACC deaminase producing Rhizobium leguminosarum RPN5 isolated from root nodules of Phaseolus vulgaris L. Bangladesh. *Journal of Botany, 45*, 477–484.

Kumar, P., Thakur, S., Dhingra, G. K., Singh, A., Pal, M. K., Harshvardhan, K., et al. (2018). Inoculation of siderophore producing rhizobacteria and their consortium for growth enhancement of wheat plant. *Biocatalysis and Agricultural Biotechnology, 15*, 264–269.

Mahé, F., de Vargas, C., Bass, D., Czech, L., Stamatakis, A., Lara, E., Dunthorn, M., et al. (2017). Parasitesdominate hyperdiverse soil protist communities in Neotropical rainforests. *Nature Ecology & Evolution, 1*(4), 1–8.

Moreira, F. M. S., & Siqueira, J. O. (2006). *Microbiologia e bioquímica do solo* (p. 729p). Universidade Federal de Lavras.

Ortiz-Hernández, M. L., Sánchez-Salinas, E., Castrejón Godínez, M. L., Dantan González, E., & Popoca Ursino, E. C. (2013). Mechanisms and strategies for pesticide biodegradation: opportunity for waste, soils and water cleaning. *Revista Internacional de Contaminación Ambiental, 29*, 85–104.

Prabhakaran, A., Meenatchi, R., Pal, S., Hassan, S., Bramhachari, P. V., Kiran, G. S., & Selvin, J. (2022). Soil microbiome: Characteristics, impact of Climate Change and resilience. In P. Veera Bramhachari (Ed.), *Understanding the microbiome interactions in agriculture and the environment*. Springer.

Prasad, S., Malav, L. C., Choudhary, J., Kannojiya, S., Kundu, M., Kumar, S., & Yadav, A. N. (2021). Soilmicrobiomes for healthy nutrient recycling. In *Current trends in microbial biotechnology forsustainable agriculture* (pp. 1–21). Springer.

Rao, M. B., Tanksale, A. M., Ghatge, M. S., & Deshpande, V. V. (1998). Molecular and biotechnological aspects of microbial proteases. *Microbiology and Molecular Biology Reviews, 62*, 597–635.

Selle, P. H., & Ravindran, V. (2007). Microbial phytase in poultry nutrition. *Animal Feed Science and Technology, 135*, 1–41.

Singh, R. (2019). Microbial biotechnology: a promising implement for sustainable agriculture. In J. S. Singh & D. P. Singh (Eds.), *New and future developments in microbial biotechnology and bioengineering* (pp. 107–114). Elsevier.

Suman, J., Rakshit, A., Ogireddy, S. D., Singh, S., Gupta, C., & Chandrakala, J. (2022). Microbiome as a key player in sustainable agriculture and human health. *Frontiers in Soil Science, 2*, 821589.

Tian, B., Zhang, C., Ye, Y., Wen, J., Wu, Y., Wang, H., et al. (2017). Beneficial traits of bacterial endophytes belonging to the core communities of the tomato root microbiome. *Agriculture, Ecosystems and Environment, 247*, 149–156.

Verma, S., Singh, A., Pradhan, S. S., Singh, R., & Singh, J. (2017). Bio-efficacy of organic formulations on crop production—A review. *International Journal of Current Microbiology and Applied Sciences, 6*, 648–665.

# Chapter 9
# Microbial Biosensors: Design, Types and Applications

## Contents



## 9.1 Introduction

The term "biosensor" refers to an analytical tool that uses certain biological processes to detect chemicals. Enzymes, receptors, antibodies, or even an entire cell, including single-celled microbes, animal and plant cells, or even tissues, can function as sensing elements. American biochemist L. C. Clark Jr. is the first person to invent a biosensor. The phrase "enzyme electrode" was first used in practice in 1962 by Clark and Lyons (Turner, 1996). To make this device, a small amount of glucose oxidase was placed on the platinum electrode's surface and then covered with cellophane. A biosensor is a device that gives quantitative or semi-quantitative analytical data corresponding to the target concentration by combining biological molecules as the recognition element with a physical transducer. Materials with biological inspiration that have been engineered to possess certain bioactive chemicals, desired physical characteristics, and programmable topography have become useful instruments for examining cell behaviour. Synthetic cell interfaces that have integrated sensing properties are

 
P. Saini, P. Mishra, *Bioprospecting of Microbial Resources for Agriculture, Environment and Bio-chemical Industry*, https://doi.org/10.1007/978-3-031-63844-2_9

among the most useful materials for determining the biophysical and biochemical interactions between cells and their surroundings (Saffioti et al., 2020). Many investigations have been carried out to build biosensors with improved accuracy and sensitivity by utilizing the tight connection and high specificity between biological molecules and target chemicals. Enzymes, receptor proteins, DNA, RNA, entire cells, and various antibody combinations are used as the recognition elements in these sensing devices. The use of certain microbes as sensing components is included in microbial biosensors, which are relevant in environmental monitoring, food safety, pollutants, poisons detection, etc. The first "microbial biosensor" was created by Divies in 1975 using *Acetobacter xylinum* and an oxygen electrode. Since then, numerous sorts of study have been carried out to increase the effectiveness of this form of biosensor (Srivastava et al., 2022). Biomedical diagnosis, illness progression tracking, drug development, food safety, and environmental monitoring are just a few of the many domains where biosensors continue to be crucial. These cutting-edge analytical instruments are compact gadgets that look for or recognize molecules using a biological recognition system (Tetyana et al., 2021). The search to develop more operative and cost-efficient environmental monitoring methods has improved with high sensitivity, rapidity, and selectivity compared to traditional low-detecting and analytical methods like capillary electrophoresis, potentiometric biosensor Quartz Crystal Microbalance (QCM), chromatographic techniques, Scanning Probe Microscopy (SPM), thermal and piezoelectric biosensors, polymerase chain reaction (PCR) techniques, and enzyme-linked immunosorbent assay (ELISA). Recent developments have led to the development of improved biological sensors that use bio-recognition elements to detect and assess parameters. These sensitive and potent biosensors that operate in the nanoparticle range are making environmental monitoring and food analysis more inexpensive and efficient. Furthermore, to create a more hospitable atmosphere, fluorescence as a newly emerged technique has been used in environmental monitoring systems with the help of green fluorescence protein (GFP) and yellow fluorescence protein (YFP), as well as deoxyribonucleic acid (DNA)-based biosensors (genosensors). The ability to identify bacterial cells and their metabolites using nano/micro biosensors, like surface-enhanced Raman spectroscopy (SERS), is another advancement. This biosensor detects microorganisms, chemicals, explosives, and disease markers using the spectroscopic analysis approach. This will examine and discuss environmental monitoring sensors, food-borne disease biosensors, and the application of genetically modified microbes for microbial biosensors Oluwaseun Adetunji et al. (2022). There are however, few researches concerning the use of biosensors in environmental monitoring and how they relate to the Sustainable Development Goals (SDGs). It is projected that biosensor technology will be a potent instrument for tracking environmental contaminants to benefit sustainable development. This chapter reviews the various kinds of biosensors, their building principles, applications, and performance.

## 9.2 Characteristic Features Microbial Biosensors

Biosensors, whose main purpose is express analysis for the detection of desired substances. Biological material is the primary "character" of the biosensor analyzer since it gives the device sensitivity to the drug being searched for. The creation of enzyme electrodes logically led to the appearance of microbial sensors. For both microbial and enzyme biosensors, the general mechanism of signal creation is similar. Either way, the biocatalytic reaction that follows the assertions about the kinetics of the enzyme occurs. The intracellular metabolic pathway that forms the basis of microbial biosensors can be schematically depicted as follows. Once within the cell, the analyte is transformed by the intracellular enzymes. As a result, the reaction products which may also be electrochemically active, are produced while the co-substrates are consumed. The immobilized cells' layer can be used to register the oxygen content, medium ionic composition, and other data. These can be employed as markers of the metabolic status of the cells and as a background for the electrochemical identification of biologically active substances.

The standard microbial biosensor's bioreceptor is a membrane or gel strip with immobilized microorganisms adhered to a transducer's working surface. Cell immobilization guarantees low biomaterial consumption (~10–100 mg), a high assay rate, and the ability to perform numerous tests without replacing the bioreceptor. The requirements of sensor stability and measurement rate (i.e., a support's diffusion properties) dictate which immobilization approach is selected (D'Souza, 2001a). In general, there are two types of immobilization techniques: active and passive. Microorganisms possess an inherent sticky property that can be utilized to immobilize them passively. On the other hand, cell fixation through chemical and physical means is assumed in active immobilization. The properties of the cells and the sort of intended experiment dictate which technique is most effective.

Major characteristics of the microbial sensors are as below

(a) The enzymes in the microbial cells are in a naturally occurring, evolutionary optimised environment that enables high stability of a variety of microbial sensors.
(b) The microbial cells can be genetically manipulated to produce recombinant organisms with specified biocatalytic capabilities.
(c) The presence of coenzymes and pathway activators in cells eliminates the need for their addition to the medium.
(d) The use of microorganisms enables the realization of sequential biochemical processes that would require artificial design in the case of an enzyme sensor; the microbial sensor's receptor can be regenerated by conditioning the growth of the cells.

## 9.3 Types of Microbial Biosensors

Types of microbial biosensors based on types of sensing mechanism.

1. **Electrochemical biosensors**
   1.1. Amperometric biosensor.
   1.2. Conductimetric.
   1.3. Potentiometric.
2. **Optical based microbial biosensors**
   2.1 Fluorescent microbial biosensor.
   2.2 Bioluminescent microbial biosensor.
3. **Other types of microbial biosensors**

## 9.4 Microbial Biosensors Design

The biological recognition element responds to the target compound in a biosensor's general design, and the transducer transforms the biological response into a detectable signal that can be measured electrochemically, optically, acoustically, mechanically, calorimetrically, electronically, or by other means. The detectable signal can then be correlated with the analyte concentration. Biosensors have been created using a variety of biological recognition components, such as cofactors, enzymes, antibodies, microbes, organelles, tissues, and cells from higher organisms (Lei et al., 2006; Liang et al., 2011). Due to their distinctive specificity and sensitivity, enzymes are the most frequently utilized recognition element among these biological elements (D'Souza, 2001b). The ability of microbes to respond to a variety of compounds, which can be utilized as the signal for sensing purposes, can be compared to a "factory" made up of multiple enzymes and cofactors/coenzymes. Even though microorganisms have non-specific metabolisms, highly selective microbial biosensors may be possible by blocking undesirable metabolic pathways, inducing desired ones, and adapting the microorganisms to a suitable substrate of interest (target) through careful cultivation conditions. Because of their variety and capacity to survive in challenging circumstances, these microbes are becoming more and more relevant as potential biosensors. Since they don't require as much expensive purification as enzymes do, they are more easily embraced by the public. Previously, the respiratory and metabolic processes of the chosen microorganisms were utilized for testing, but more recently, lux, gfp, or lacZ gene reporters have been fused with inducible gene promoters for extensive testing of toxicity and bioavailability in the environment. The microbial cells, which act as the biosensor's recognition element, must be closely linked to the transducer to efficiently convert the biochemical reaction into a physical signal (Fig. 9.1).

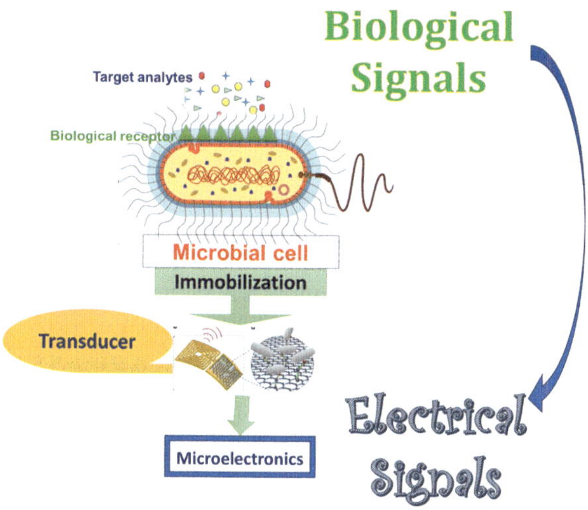


**Fig. 9.1** Pictorial representation of microbial sensor design

Immobilized viable or non-viable microbial cells combined with a transducer form a microbial biosensor. The majority of the time, less expensive alternatives to enzymes have been entire cells harboring periplasmic enzymes or non-viable cells recovered following permeabilization. To create microbial biosensors, immobilizing microorganisms on transducers is crucial (D'Souza, 2001b). Adsorption, encapsulation, entrapment, covalent binding, and cross-linking are examples of conventional techniques for immobilizing microorganisms.

## 9.5 Potential Applications and Utilization of Microbial Biosensors

Recently, genetically altered microbes have become more and more important in strengthening the effectiveness of biosensors. Microbial biosensors, which are important in environmental monitoring, food safety, pollutants, toxins detection, etc., use specific bacteria as sensing components. The choice of microbes and their immobilization on the transducer via chemical and physical techniques are crucial stages that need to be taken for the applications of biosensors. Some of the primary uses are biological oxygen demand (BOD) detection, surfactant detection, metal ion detection, inorganic acid detection, hydrocarbon detection, genotoxicity etc. Numerous microbial biosensors have been created recently for use in food, medicine, and the environment. Some potential applications are given below:

### 9.5.1 *Environmental Monitoring*

Human health is greatly at danger from environmental pollutants. Many microbial biosensors for the identification of both inorganic and organic toxins. An inexpensive, targeted, easy-to-use, and fast tool is required for heavy metal monitoring. The microbial biosensor offers a chance to resolve this issue. Two general approaches for creating a microbial biosensor to track heavy metal toxicity are the constitutive technique (light-off) and the inducible manner (light-on) (Ivask et al., 2009). More advanced and sensitive biosensors are available; these biosensors rely on inducible promoters coupled to reporter genes (Xu & Ying, 2011). A microbial biosensor for the detection of Pb (II) was created using heat-dried *Pseudomonas aeruginosa*, yielding a linear response to Pb (II) from 1.0 μM to 2.0 μM with a detection limit of 0.6 μM (Yüce et al., 2011).

### 9.5.2 *Food Processing and Fermentation*

Rapid and inexpensive ways to guarantee product quality and process controls are necessary since the government and customers both have high standards for the quality of the products (Mello & Kubota, 2002). Both consumers and the food business are becoming more interested in controlling the quality and freshness of food (Mello & Kubota, 2002). To monitor nutritional parameters and food pollutants, there is a need for analytical methods that are both specific and fast (D'Souza, 2001a). Microbial biosensors are an efficient and reasonably priced way to ensure product quality. The two best sweeteners and nutritional additives that are frequently used in food are D-glucose and D-xylose. A voltametric biosensor for the detection of D-glucose and D-xylose was created based on the co-immobilization of glucose oxidase and xylose dehydrogenase-displayed XDH-bacteria on multiwalled CNT nanocomposite films modified electrode (Li et al., 2013). Furthermore, genetically engineered *Saccharomyces cerevisiae* strains were employed as the bio-element of the microbial biosensor for the detection of zearalenone family mycotoxins in milk to ensure the quality of the milk (Valimaa et al., 2010). Foods and beverages are produced via fermentation, which calls for a meticulously managed fermentation system operation (Kim et al., 2009). To regulate the fermentation process, microbial biosensors are utilized to monitor the ingredients. To assess ethanol during the fermentation process in flow injection analysis, Valach et al. (2009) built a novel microbial amperometric biosensor that, in 3 min, demonstrated a linear response to ethanol in the range of 10 μM to 1.5 mM.

### 9.5.3 *Medical and Health Diagnosis*

Microbial biosensors offer a quick, precise, and affordable method of diagnosing diseases, hormones, and DNA—all significant aspects of a living being. A microbial biosensor for pathogen detection was created using mast cells from rats with basophilic leukaemia (RBL), which could demonstrate a strong exocytotic response in a matter of minutes with the addition of an antigen (Curtis et al., 2008).

## 9.6 Environmental Sustainability of Microbial Biosensors

The state of the environment worldwide, especially climate change brought on by global warming, is getting worse every year. The United Nations Sustainable Development Summit established "The 2030 Agenda" and "The Sustainable Development Goals (SDGs)" in 2015 because we have to change the planet if we hope for human survival. There are various applications of biosensorsfor environmental monitoring discussed as follows

- *Heavy metals monitoring*: Heavy metals may be readily detected by biosensors, which allows for the management of water quality and safety.
- *Organic pollutants monitoring*: organic pollutants including polychlorinated biphenyls (PCBs), polybrominated biphenyls (PBBs), polybrominated diphenyl ethers (PBDEs), phthalate esters (PAEs), and other hazardous pollutants exist in industrial wastewater may be detected by biosensors
- *Monitoring of wastewater quality*: Biosensors, particularly those based on Microbial fuel cell (MFC), have been used worldwide for wastewater monitoring

Microbial biosensors can help achieve a number of the objectives listed in this agenda, including "conservation of water and soil resources" and "preservation of food safety." These tissue- or cell-based biosensors now have unique and unrealized benefits thanks to state-of-the-art advancements (Liu et al., 2014). While molecular biosensors are more sensitive to environmental signals than whole-cell biosensors (WCBs), WCBs can perceive various responses within a living cell through modification through straightforward genetic engineering techniques. Only living cells can be used to evaluate crucial environmental factors as bioavailability, cytotoxicity, and mutagenicity or genotoxicity. Due to their associated benefits of enhanced analytical selectivity, rapid, reliable, sensitive, and cost-efficient monitoring with limit of detection (LOD), whole-cell microbial-derived biosensors appear to be essential for the efficient detection and sensing of environmentally related toxic compounds or contaminants (Maduraiveeran & Jin, 2017). Furthermore, microbially generated biosensors are affordable, portable, small, and simple to operate. Their distinct structural and functional characteristics make them very appropriate for the identification and tracking of various contaminants associated with the environment.

## 9.7 Conclusion

Because of the state-of-the-art advancements made in environmental detection and biological diagnostics over the past 20 years, microbial-derived biosensors have drawn a lot of attention from researchers. Unlike conventional chemically-based sensors, whole-cell derived biosensors were able to quickly, accurately, and in real time identify a wide range of different contaminants or dangerous substances from a wider spectrum.Good environmental management and monitoring require the ability to swiftly identify and provide information on the contamination zone and dangerous chemicals, which biosensors can do.Recent researchhas proposed the performance and application of biosensors in genuine environmental samples, including soil, wastewater, and river samples. Further investigation into the actual environmental sample tests is required for improved commercial and environmental feasibility.

## References

Curtis, T., Naal, R. M. Z. G., Batt, C., Tabb, J., & Holowka, D. (2008). Development of a mast cell-based biosensor. *Biosensors and Bioelectronics, 23*(7), 1024–1031.

D'Souza, S. F. (2001a). Immobilization and stabilization of biomaterials for biosensor applications. *Applied Biochemistry and Biotechnology, 96*(1–3), 225–238.

D'Souza, S. F. (2001b). Microbial biosensors. *Biosensors and Bioelectronics, 16*(6), 337–353.

Ivask, A., Rolova, T., & Kahru, A. (2009). A suite of recombinant luminescent bacterial strains for the quantification of bioavailable heavy metals and toxicity testing. *BMC Biotechnology, 9*, 41–55.

Kim, Y., Park, J., & Jung, H. (2009). An Impedimetric biosensor for real-time monitoring of bacterial growth in a microbial Fermentor. *Sensor & Actuators B: Chemical, 138*(1), 270–277.

Lei, Y., Chen, W., & Mulchandani, A. (2006). Microbial biosensors. *Anal Chim Acta, 568*, 200–210. https://doi.org/10.1016/j.aca.2005.11.065

Li, L., Liang, B., Li, F., Shi, J., Mascini, M., Lang, Q., & Liu, A. (2013). Co-immobilization of glucose oxidase and xylose dehydrogenase displayed whole cell on multiwalled carbon nanotube nanocomposite films modified electrode for simultaneous voltammetric detection of d-glucose and d-xylose. *Biosensors and Bioelectronics, 42*(12), 156–162.

Liang, S., Jia, W., Hou, C., & Lei, Y. (2011). Microbial biosensors: A review. *Biosensors and Bioelectronics, 26*(5), 1788–1799.

Liu, B., Chen, W., Evavold, B. D., & Zhu, C. (2014). Accumulation of dynamic catch bonds between TCR and agonist peptide-MHC triggers T cell signaling. *Cell, 157*, 357–368.

Maduraiveeran, G., & Jin, W. (2017). Nanomaterials based electrochemical sensor and biosensor platforms for environmental applications. *Trends in Environmental Analytical Chemistry, 13*, 10–23.

Mello, L. D., & Kubota, L. T. (2002). Review of the use of biosensors as analytical tools in the food and drink industries. *Food Chemistry, 77*(2), 237–256.

Oluwaseun Adetunji, C., Mathew, J. T., Inobeme, A., Olaniyan, O. T., Singh, K. R. B., Abimbola, O. F., Nayak Singh, V. J., & Singh, R. P. (2022). Microbial and plant cell biosensors for environmental monitoring. In R. P. Singh, K. E. Ukhurebor, J. Singh, C. O. Adetunji, & K. R. Singh (Eds.), *Nanobiosensors for environmental monitoring*. Springer. https://doi.org/10.1007/978-3-031-16106-3_9

Saffioti, N. A., Cavalcanti-Adam, E. A., & Pallarola, D. (2020). Biosensors for studies on adhesion-mediated cellular responses to their microenvironment. *Frontiers in Bioengineering and Biotechnology, 8*, 597950.

Srivastava, P., Prasad, D., & Nigam, V. K. (2022). Chapter 19: Insight into microbial biosensors: Design, types and applications. In P. Verma & M. P. Shah (Eds.), *Bioprospecting of microbial diversity* (pp. 425–440). Elsevier.

Tetyana, P., Shumbula, P. M., & Njengele-Tetyana, Z. (2021). *Biosensors: Design, development and applications*. Nanopores [Working Title].

Turner A. P. F. (1996). *Biosensors: Past, present and future*. http://www.cranfield.ac.uk/health/researchareas/biosensorsdiagnostics/page18795.jsp

Valach, M., Katrlik, J., Sturdik, E., & Gemeiner, P. (2009). Ethanol Gluconobacter biosensor designed for flow injection analysis: Application in ethanol fermentation off-line monitoring. *Sensors & Actuators B: Chemical, 138*(2), 581–586.

Valimaa, A., Kivisto, A. T., Leskinen, P. I., & KarpM, T. (2010). A novel biosensor for the detection of Zearalenone family mycotoxins in Milk. *Journal of Microbiological Methods, 80*(1), 44–48.

Xu, X., & Ying, Y. (2011). Microbial biosensors for environmental monitoring and food analysis. *Food Reviews International, 27*(3), 300–329.

Yüce, M., Nazir, H., & Dönmez, G. (2011). Utilization of heat dried Pseudomonas aeruginosa biomass for voltammetric determination of Pb (II). *New Biotechnology, 28*(4), 356–361.

# Chapter 10
# Role of Bioinformatics Tools in Microbial Prospective and Its Future

**Contents**



## 10.1 Introduction

The application of computer and analysis tools to the collection and analysis of biological data is referred to as bioinformatics. It is a multidisciplinary field that integrates biology, computer science, and statistics (Fig. 10.1). Expert bioinformaticians today employ sophisticated software programs to retrieve, classify, analyze, predict, and store DNA and protein sequence data as the field of bioinformatics continues to advance. Responses to communicable illnesses in clinical and public health are being revolutionised by microbial genomics. Bioinformatics advances are making it possible to respond to disease outbreaks precisely and quickly. Bioinformatic tools are widely utilized to manage the large amounts of genomic and proteomic data involved in DNA/protein sequence management, drug design, homology modelling, motif/domain prediction, docking, annotation, and dynamic simulation, among other tasks. Several different professions have incorporated bioinformatics, which helps to speed up the process of better understanding data (Fulekar, 2009). Bioinformatics is expanding quickly and offering new methods and

 
P. Saini, P. Mishra, *Bioprospecting of Microbial Resources for Agriculture, Environment and Bio-chemical Industry*, https://doi.org/10.1007/978-3-031-63844-2_10

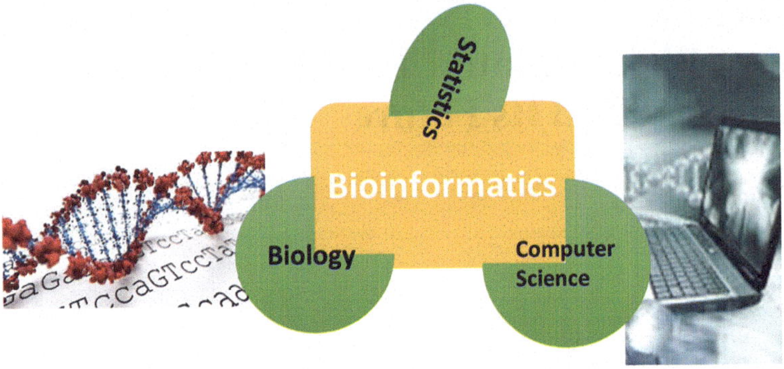


**Fig. 10.1** Bioinformatics as multidisciplinary approach

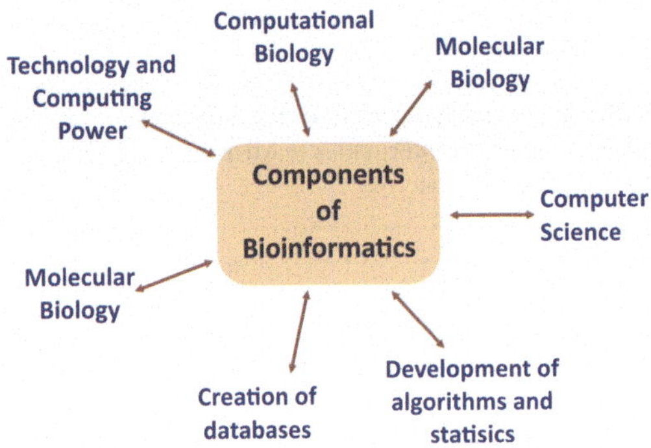


**Fig. 10.2** Components of bioinformatics

approaches for the evaluation of useful data as a result of the tremendous advancement in information technology. The collection, storage, cataloguing, and analysis of data in a special format that is simple to alter for future researches are all key aspects of bioinformatics (Gu & Bourne, 2009). Data mining and manipulations is an important aspect of bioinformatics (Piatetsky-Shapiro, 1996).

## 10.2 Characteristic Features of Bioinformatics Tools

A bioinformatician's primary resources are the internet and computer applications. Sequence analysis of DNA and proteins utilizing various online databases and programs is a basic activity. The study of the creation and use of computational tools for the analysis and interpretation of biological data is the focus of the quickly expanding discipline of bioinformatics in biology. The major components of bioinformatics are depicted in Fig. 10.2.

Biological data is processed, stored, analyzed, and interpreted using databases, tools, and bioinformatics software (Table 10.1).

**Table 10.1** List of bioinformatic sources and websites

| S. N. | Bioinformatic source | websites | Information/function |
|---|---|---|---|
| 1. | National Center for Genome Resources | www.ncgr.org/ | links scientists to bioinformatics solutions by collaborations, data, and software development |
| 2. | National Center for Genome Resources | www.ncgr.org/ | links scientists to bioinformatics solutions by collaborations, data, and software development |
| 3. | National Center for Biotechnology Information | www.ncbi.nlm.nih.gov | maintains bioinformatic tools and databases |
| 4. | European Bioinformatics Institute | www.ebi.ac.uk | centre for research and services in bioinformatics; manages databases of biological data |
| 5. | Ensembl | www.ensembl.org | automatic annotation database on genomes |
| 6. | International Society for Computational Biology | www.iscb.org/ | aims to advance scientific understanding of living systems through computation; has useful bioinformatic links |
| 7. | SWISS-PROT () | www.expasy.org/sprot/ | important protein database with sequence data from all organisms, which has a high level of annotation (includes function, structure, and variations) and is minimally redundant (few duplicate copies |
| 8. | BioInform | www.bioinform.com | global bioinformatics news service |
| 9. | Genbank | www.ncbi.nlm.nih.gov/Genbank | stores and archives DNA sequences from both large scale genome projects and individual laboratories |
| 10. | Unigene | www.ncbi.nlm.nih.gov/UniGene | gene sequence collection containing data on map location of genes in chromosomes |

Software programs known as "bioinformatic tools" are made to do sequence or structural analysis and to extract useful information from the vast amounts of biological and molecular biology databases

Bioinformatic tools are needed to maintain and analyze the DNA sequences from various organisms to handle the growing volume of genomic data (Yang et al., 2008). Utilizing various bioinformatics tools and software programs enables the determination of sequence homology, gene discovery, coding region identification, structural and functional analysis of genomic sequences, etc. (Sensen, 2005). A list of a few bioinformatics tools used in genomics and proteomics, is provided below.

- Carrie – Transcriptional regulatory networks database (Haverty & Weng, 2004)
- CisML – Motif detection tool (Haverty et al., 2004)
- ICSF -Identify conserved structural features in TF binding sites (Burden & Weng, 2005)
- Possum – Motif searching tool (Fu et al., 2004)
- Promoser – Promoter extraction tool from eukaryotic organisms (Halees et al., 2003)

- REPFIND – Determination of clustered repeats in DNA fragments (Hamilton & Davis, 2007)
- Cluster – Buster-Motifs cluster in DNA sequences prediction tool (Frith et al., 2003)
- Cister – Determines regulatory regions in DNA fragments (Kessler et al., 1993)
- Clover – Find overrepresented motifs in DNA sequences (Frith et al., 2004)
- MotifViz – Identify overrepresented motifs (Fu et al., 2004)
- RANKGENE-Gene expression datatool (Su et al., 2003)
- ROVER – Predicts overrepresented motifs in DNA fragments (Colaert et al., 2010)
- SeqVISTA-Tool for sequences viewing (Hu et al., 2003)
- Tractor-A method to identify transcription factors that have an excess of binding sites in the upstream sections of human genes that are co-expressed (Gang-tiana et al., 2007)
- OHMICS-Integrated Computational system with oral human microbiome (Tamburini et al., 2016)
- BLAST-Sequencealignment tool for DNA or protein (Moreno-Hagelsieb & Latimer, 2007)
- HMMER-Searching tool for homologous protein sequences (Finn et al., 2011)
- Clustal Omega – Multiple sequence alignments tool (Sievers & Higgins, 2014)
- Sequerome – Sequence profiling tool (Dinodia, 2015)
- ORF Finder – open reading frame in putative genes (Rombel et al., 2002; Tatusov & Tatusov, 2007)
- Virtual Footprint-Whole prokaryotic genome analysis (Münch et al., 2005)
- MEGA-To study evolutionary relationship (Sohpal et al., 2010)
- MOLPHY – Maximum likelihood-based phylogenetic analysis tool (Mehmood et al., 2014)
- PHYLIP-Phylogenetic studytool (Plotree & Plotgram, 1989)
- K2 / FAST – Protein structure alignment tool (Fogel & Corne, 2002)
- SMM – Tool for determining peptides binding to major histocompatibility complex (Chen et al., 2011)
- ZDOCK – Protein-protein docking tool (Pierce et al., 2014)
- Docking Benchmark – Docking algorithms performance tool (Chen et al., 2003)
- ZDOCK Server – An automated server for running ZDOCK (Pierce et al., 2014)
- Z3OnWeb.com – Analyzing 2D-Gel images ofproteomic analysis (Smith, 2000)

## 10.3 Microbial Prospective of Bioinformatics

Microbes are present in practically every habitat and hence, bioengineering, medical health, and industrial fermentation are important as commercial applications. In 2020, microbial bioinformatics will continue to be a dynamic and innovative field that adds value to the constantly increasing volume of new sequence data while adopting new technology and methods (Pallen, 2016). We have a significant understanding of the genetic composition of the microbial world which is possible due to genome sequencing. It is evident that microbial taxonomy is already broken and cannot adapt to a future where the majority of microorganisms will be identified and characterised by the analysis of macromolecular sequences. This is because microbial taxonomy uses a polyphasic approach that necessitates laboratory culture and phenotypic investigation. The goal of synthetic biology is to go beyond reading and write DNA sequences, and this will present both new opportunities and challenges. These obstacles may arise in the form of novel techniques to data management and storage or in the production of synthetic microbes. Microbial bioinformatics is moving away from the traditional self-managed server or cluster operated by a single user or research group in terms of hardware and software provisioning (Pallen, 2016). Bioinformatics helps to better understand the effects that bacteria have on people and the environment so that they can use them more effectively. By defining the human microbiome at key body regions, metagenomics and 16S rRNA sequencing can reveal the relationships between microbes and human health. The application of next-generation sequencing (NGS) technology to metagenomic sequencing has made significant strides in the study of the human microbiome.

The organic waste from breweries, textile, agro-based, residential wastewater systems, and landfills is used to create microbial fuel cells (MFC). MFC may also be used to produce energy, lower the need for chemical oxygen, and aid in the degradation of different waste products. Bioinformatics methodologies are being frequently applied in the technology of MFCdue to the rapid progress of sequencing techniques. They also have a significant impact on the composition of electroactive species found in microbes. Bioinformatics resources can also be used to determine the metabolic route.

The fields of current bioinformatics research include: (i) genomics, which studies genes and genome functionality through sequencing and comparative analysis; (ii) proteomics, which identifies and characterises protein-related properties and reconstructs metabolic and regulatory pathways; (iii) cell visualisation and simulation, which models and studies cell behaviour; and (iv) application to drug and antimicrobial agent development. Microbes have been studied at a very basic level with the use of bioinformatics techniques required to analyze the unique gene set that enables them to survive in hostile environments (Kerkhoven et al., 2004). Microbial bioprospecting is the most researched and contested kind of bioprospecting out of all of the others. The last 10 years have seen a significant increase in the use of genome mining bioinformatics techniques for microbial bioprospecting.

## 10.4 Role of Bioinformatics in Microbiology

A vast range of applications in many fields, including genetics, are provided by bioinformatics. Forensics, proteomics, comparative genomics, nutrigenomics, microbial genome, etc. The analysis of gene variation and expression, the analysis and prediction of gene and protein structure and function, the prediction and detection of gene regulatory networks, simulation environments for whole cell modelling, complex modelling of gene regulatory dynamics and networks, and the presentation and analysis of molecular pathways are just a few of the important tasks that bioinformatics is now being used for (Tsoka & Ouzounis, 2000). In Microbiology bioinformatics enables recognition of strains and species, describing the core- and pan-genomes the shared and unique genes of various strains, connecting the functional and genomic components of strains to growth circumstances or clinical factors, andfinding new genes, especially those related to enzymes, for use in industry or medicine. Moreover, for a deeper understanding of microbes and their environments, as well as for many other applications of microbiology, such as barcoding, microbial bio-remediation, and the production of biofuel, data analysis is crucial (Bishop, 2014). These applications range from understanding the underlying causes of diseases or host-pathogen interactions to designing medications and developing vaccines. The topics of food safety are given more consideration, particularly the usage of genetically modified foods and the safety of microorganisms and biology (Wani et al., 2018). Therefore, accelerating scientific research in biotechnology offers a bright future.

## 10.5 Future Prospective

Finding cellular networks of genes, comprehending their intricate interactions, and determining their function in disease are replacing the practice of examining single genes in isolation to explain hereditary disorders (Debouk & Metcalf, 2000). This will bring forth a completely new era of personalized medicine. The benefits of computational biology will be tapped into by molecular biologists and clinical researchers with the help of bioinformatics (Butler, 2001). Clinical research teams who can seamlessly transition between the laboratory bench, clinical practice, and the use of these advanced computational tools will be the most effective in the future decades.

## 10.6 Conclusion

Numerous novel discoveries have been made as a result of research into various microbiomes, including genes that encode proteins that could have major commercial significance. The group of microorganisms that inhabit a specific ecosystem is known as the microbiome, and metagenomics is the technique of concurrently sequencing the genomes of all the organisms in a given biological sample. With the use of bioinformatics and software technologies, researchers and regulators have been driven to develop and explore new and complex approaches to evaluate promising new goods. This has resulted in trial designs that are more targeted, flexible, and rapid. Thanks to technological advancements in sequencing, metagenomics has enabled us to assess microbiomes at a depth never before possible. Microbial bioinformatics will continue to be a dynamic and innovative field that embraces new technology and methods while contributing to the constantly increasing volume of fresh sequence data. Microbial sequence space will continue to be the last frontier as we venture into the third decade of the twenty-first century.

## References


Bishop, O. T. (2014). *Bioinformatics and data analysis in microbiology*. Caister Academic Press, Rhodes University.

Burden, H. E., & Weng, Z. (2005). Identification of conserved structural features at sequentially degenerate locations in transcription factor binding sites. *Genome Informatics, 16*(1), 49–58.

Butler, D. (2001). Are you ready for the revolution? *Nature, 409*, 758–760.

Chen, R., Mintseris, J., Janin, J., & Weng, Z. (2003). A protein–protein docking benchmark. *Proteins: Structure, Function, and Bioinformatics, 52*(1), 88–91.

Chen, P., Rayner, S., & Hu, K. H. (2011). Advances of bioinformatics tools applied in virus epitopes prediction. *Virologica Sinica, 26*(1), 1–7.

Colaert, N., Helsens, K., Impens, F., Vandekerckhove, J., & Gevaert, K. (2010). Rover: A tool to visualize and validate quantitative proteomics data from different sources. *Proteomics, 10*(6), 1226–1229.

Debouk, C., & Metcalf, B. (2000). The impact of genomics on drug discovery. *Annual Review of Pharmacology and Toxicology, 40*, 193–208.

Dinodia, A. (2015). Bioinformatics–Emerging area in medical and biomedical informatics. *International Journal of Pharmaceutical Chemistry and Analysis, 2*(4), 187–191.

Finn, R. D., Clements, J., & Eddy, S. R. (2011). HMMER web server: Interactive sequence similarity searching. *Nucleic Acids Research, 39*(suppl_2), W29–W37.

Fogel, G. B., & Corne, D. W. (Eds.). (2002). *Evolutionary computation in bioinformatics*. Morgan Kaufmann.

Frith, M. C., Li, M. C., & Weng, Z. (2003). Cluster-Buster: Finding dense clusters of motifs in DNA sequences. *Nucleic Acids Research, 31*(13), 3666–3668.

Frith, M. C., Fu, Y., Yu, L., Chen, J. F., Hansen, U., & Weng, Z. (2004). Detection of functional DNA motifs via statistical over-representation. *Nucleic Acids Research, 32*(4), 1372–1381.

Fu, Y., Frith, M. C., Haverty, P. M., & Weng, Z. (2004). MotifViz: An analysis and visualization tool for motif discovery. *Nucleic Acids Research, 32*(suppl_2), W420–W423.

Fulekar, M. H. (2009). *Bioinformatics: Applications in life and environmental sciences*. Springer.

Gang-tiana, L. I. U., Bai-ningb, G. A. O., & Hui-mina, C. A. O. (2007). Realization of tractor product design management based-on network technique. *Tractor & Farm Transporter, 4*, 043.

Gu, J., & Bourne, P. E. (2009). *Structural bioinformatics* (Vol. 44). Wiley.

Halees, A. S., Leyfer, D., & Weng, Z. (2003). PromoSer: A large-scale mammalian promoter and transcription start site identification service. *Nucleic Acids Research, 31*(13), 3554–3559.

Hamilton, R. S., & Davis, I. (2007). RNA localization signals: Deciphering the message with bioinformatics. In *Seminars in cell & developmental biology* (Vol. 18 (2), pp. 178–185). Academic.

Haverty, P. M., & Weng, Z. (2004). CisML: An XML-based format for sequence motif detection software. *Bioinformatics, 20*(11), 1815–1817.

Haverty, P. M., Frith, M. C., & Weng, Z. (2004). CARRIE web service: Automated transcriptional regulatory network inference and interactive analysis. *Nucleic Acids Research, 32*(suppl_2), W213–W216.

Hu, Z., Frith, M., Niu, T., & Weng, Z. (2003). SeqVISTA: A graphical tool for sequence feature visualization and comparison. *BMC Bioinformatics, 4*(1), 1.

Kerkhoven, R., Van Enckevort, F. H., Boekhorst, J., Molenaar, D., & Siezen, R. J. (2004). Visualization for genomics: The microbial genome viewer. *Bioinformatics, 20*, 1812–1814.

Kessler, L., Fintor, L., Muha, C., Wun, L. M., Annett, D., & Mazan, K. D. (1993). The Cancer Information Service Telephone Evaluation and Reporting System (CISTERS): A new tool for assessing quality assurance. Journal of the National Cancer Institute. *Monograph, 14*, 61–65.

Mehmood, M. A., Sehar, U., & Ahmad, N. (2014). Use of bioinformatics tools in different spheres of life sciences. *Journal of Data Mining in Genomics & Proteomics, 5*(2), 1.

Moreno-Hagelsieb, G., & Latimer, K. (2007). Choosing BLAST options for better detection of orthologs as reciprocal best hits. *Bioinformatics, 24*(3), 319–324.

Münch, R., Hiller, K., Grote, A., Scheer, M., Klein, J., Schobert, M., & Jahn, D. (2005). Virtual Footprint and PRODORIC: An integrative framework for regulon prediction in prokaryotes. *Bioinformatics, 21*(22), 4187–4189.

Pallen, M. J. (2016). Microbial bioinformatics 2020. *Microbial Biotechnology, 9*(5), 681–686.

Piatetsky-Shapiro, G. (1996). In U. M. Fayyad, P. Smyth, & R. Uthurusamy (Eds.), *Advances in knowledge discovery and data mining* (p. 21). AAAI Press.

Pierce, B. G., Wiehe, K., Hwang, H., Kim, B. H., Vreven, T., & Weng, Z. (2014). ZDOCK server: Interactive docking prediction of protein–protein complexes and symmetric multimers. *Bioinformatics, 30*(12), 1771–1773.

Plotree, D. O. T. R. E. E., & Plotgram, D. O. T. G. R. A. M. (1989). PHYLIP-phylogeny inference package (version 3.2). *Cladistics, 5*(163), 6.

Rombel, I. T., Sykes, K. F., Rayner, S., & Johnston, S. A. (2002). ORF-FINDER: A vector forhigh-throughput gene identification. *Gene, 282*(1), 33–41.

Sensen, C. (2005). *Handbook of genome research: Genomics, proteomics, metabolomics, bioinformatics, ethical and legal issues*. Wiley.

Sievers, F., & Higgins, D. G. (2014). Clustal Omega, accurate alignment of very large numbers of sequences. *Multiple Sequence Alignment Methods*, 105–116.

Smith, C. M. (2000). Bioinformatics, genomics, and proteomics. *The Scientist, 14*(23), 26–26.

Sohpal, V. K., Dey, A., & Singh, A. (2010). MEGA biocentric software for sequence and phylogenetic analysis: A review. *International Journal of Bioinformatics Research and Applications, 6*(3), 230–240.

Su, Y., Murali, T. M., Pavlovic, V., Schaffer, M., & Kasif, S. (2003). RankGene: Identification of diagnostic genes based on expression data. *Bioinformatics, 19*(12), 1578–1579.

Tamburini, S., Shen, N., Wu, H. C., & Clemente, J. C. (2016). The microbiome in early life: Implications for health outcomes. *Nature Medicine, 22*(7), 713–722.

Tatusov, T., & Tatusov, R. (2007). *ORF Finder (Open Reading Frame Finder)* (p. 9). National Center for Biotechnology Information. National Institute of Health.

Tsoka, S., & Ouzounis, C. A. (2000). Recent developments and future directions in computational genomics. *FEBS Letters, 480*, 42–48.

Wani, M. Y., Ganie, N. A., Rani, S., Mehraj, S., Mir, M. R., Baqual, M. F., Sahaf, K. A., Malik, F. A., & Dar, K. A. (2018). Advances and applications of bioinformatics in various fields of life. *International Journal of Fauna and Biological Studies, 5*(2), 03–10.

Yang, J. Y., Yang, M. Q., Zhu, M. M., Arabnia, H. R., & Deng, Y. (2008). Promoting synergistic research and education in genomics and bioinformatics. *BMC Genomics, 9*(1), I1.

# Chapter 11
# Revolution in Microbial Bioprospecting via the Development of Omics-Based Technologies

**Contents**



## 11.1 Introduction

Microorganisms, constituting the majority of life on Earth, display remarkable biodiversity and metabolic adaptability across diverse environments. Microbial biodiversity surpasses that of eukaryotes, making the microbial realm a valuable and seemingly limitless source for discovering novel bioactive compounds. Microbial communities have a close connection with plants and animals, providing essential nutrients, metals, and vitamins to their hosts. In the case of humans, the vast




P. Saini, P. Mishra, *Bioprospecting of Microbial Resources for Agriculture, Environment and Bio-chemical Industry*, https://doi.org/10.1007/978-3-031-63844-2_11

population of gut microbes plays a crucial role in aiding digestion, metabolizing toxins, and defending against pathogens. The significance of microbes extends beyond nutrition and health, encompassing environmental benefits such as their role in cleaning up pollutants like oil and chemical spills (NRCCM, 2007). Their ability to thrive in extreme conditions underscores their omnipresence.

The genomes of microbes encompass a diverse range of biochemical transformations, reflecting the accumulated DNA changes shaped by billions of years of environmental changes and evolution (Handelsman, 2004). Human civilization has witnessed significant advancements attributable to various technologies rooted in microbial contributions. Microbes play a pivotal role in the production of antibiotics and drugs for clinical applications, the remediation of soil and water pollutants, the generation of biofuels, the improvement and protection of agricultural crops, and the fermentation of human foods. Additionally, microbes serve as markers for disease detection, showcasing their multifaceted impact on enhancing various aspects of human life. Despite their potential, more than 99% of microorganisms remain unexplored due to the limitations of culture-dependent methods, which only support the growth of a fraction of microbial species. To overcome this constraint, culture-independent techniques have emerged, revolutionizing the study of microbial diversity and community dynamics. Genomic and metagenomic approaches, characterized by high throughput capabilities, have accelerated the exploration of previously deemed "uncultivable" microorganisms (Bansal et al., 2022). The transformation in understanding the microbial world has been driven by the introduction of genome-based techniques that utilize various omic approaches. These advanced methods, characterized by their high-throughput capabilities, enable the investigation of previously inaccessible microbial life forms in significantly reduced time periods. This has greatly enhanced the effectiveness and efficiency of microbial bioprospecting endeavors.

The revolutionary advancements in omics have played a crucial role in unraveling the physiological and biochemical characteristics of diverse organisms. This includes commercially significant organisms that contribute to various industrial processes. The adoption of highly efficient whole genome-sequencing strategies has facilitated a comprehensive comprehension of the functional annotations associated with various gene products (Golden & Handfield, 2014). To attain a complete understanding of microbial biological processes, it becomes imperative to quantify them through functional characterization studies involving genes, coding regions, proteins, and metabolic products. The progress in these domains has given rise to a new set of technical expertise known as genomics, transcriptomics, proteomics, metabolomics, and interactomics (Low et al., 2013). These approaches are characterized by their high efficiency and reliance on high-throughput mechanisms. Through the integration of experimental methodologies with various computational and statistical tools, substantial amounts of highly representative data can be generated (Berger et al., 2013). It is important to note that, given the rationale, no single high-throughput omics approach can offer a comprehensive understanding of metabolic processes. Hence, multiple approaches must be employed and coordinated to achieve industrial targets. The utilization of these omics approaches is now considered fundamental to any industrially viable procedure. Industrial processes heavily depend on the development and accessibility of highly efficient microbial strains to ensure the maximum production of various bioproducts (Keasling, 2012). Omics approaches not only contribute to a comprehensive understanding of biological processes but also play a key role in the creation of microbial entities. To enhance the overall efficiency of omics approaches in the industry, experimental procedures must be optimized in tandem with computational and statistical models. These technologies offer distinct advantages in exploring microbial diversity, adaptive evolution, alterations in metabolic characteristics, and the resistance mechanisms of individual strains or their communities in extreme environments. Moreover, omics technologies are valuable tools for discovering potential new genes, enzymes, metabolites, metabolic pathways, and species. The integration of multi-omics analysis enables the linking of information across different biomolecular levels, providing a more accurate and complete understanding of the global adaptation mechanisms of microorganisms. The current chapter provides a review of the importance of omics tools in the industrial biotechnology processes. The different domains within omics technology and their significance have then been summarized. The chapter also addresses the challenges and future perspectives in adapting omics technology for microbial bioprospecting.

## 11.2 Significance of Omics-Based Technologies

Emerging technologies enabling the simultaneous monitoring of numerous macro- and small molecules offer the prospect of functionally assessing multiple, if not all, vital cellular pathways concurrently. These innovative "global" approaches, encompassing the analysis of entire families of cellular molecules such as RNA, proteins, and intermediary metabolites, are collectively referred to as "-omic" technologies. This nomenclature stems from their capability to comprehensively characterize all or most members of a molecular family in a single analysis. With the aid of these advanced tools, we can now achieve comprehensive evaluations of the functional dynamics of biochemical pathways and identify structural genetic variations (sequence differences) among individuals and species, surpassing previous limitations (Debnath et al., 2010). Omics technologies, also known as high-dimensional biology, encompass a comprehensive study of cells, tissues, and organisms by integrating data from various platforms to facilitate interpretation. These technologies primarily identify genes (genomics), mRNAs (transcriptomics), proteins (proteomics), and metabolites (metabolomics) in an unbiased, non-targeted manner (Fig. 11.1).

The exploration of integration and interrelationships among networks of biological processes falls under the domain of systems biology. This holistic approach allows for a more thorough understanding of the intricate connections within biological systems. Omic technologies enable the comprehensive analysis of transcription, translation, and metabolism at the molecular level (Xu et al., 2018; Amer &

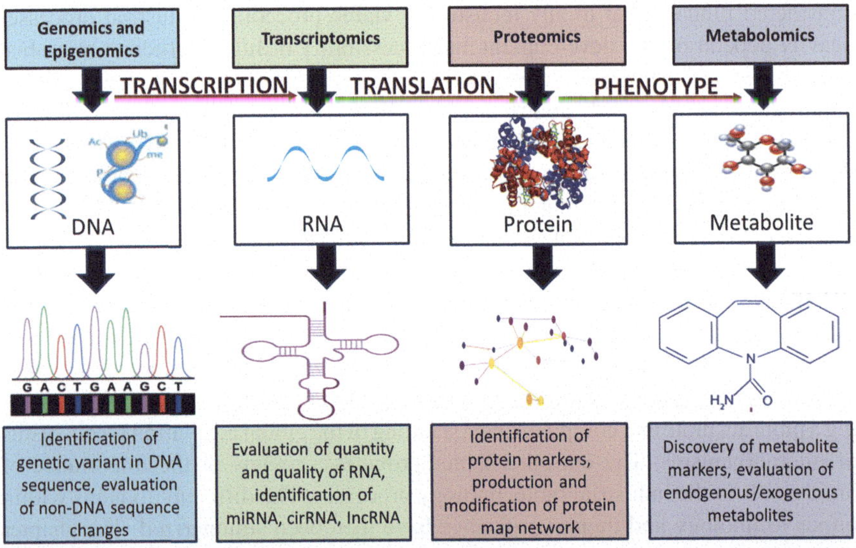


**Fig. 11.1** Omics technologies in microbial bioprospecting

Baidoo, 2021). Through omics analysis of diverse microbial phenotypes, a deeper understanding of overall metabolism changes in response to microenvironmental shifts is achieved. This insight into macro factors affecting microbial performance spans the molecular cell to microenvironment, facilitating precise and swift optimization. The capacity to simultaneously detect a significantly higher number of proteins/genes enhances the ability to correlate complex mixtures with intricate effects, manifested as gene/protein expression profiles. The rapid integration of omic data in pharmaceutical and biotechnology research is giving rise to commercial applications, with organizations incorporating various omic datasets into comprehensive computer models for systems biology solutions. This convergence promises to provide a more holistic understanding of complex biological systems.

## 11.3 Omics Technologies in Microbial Bioprospecting

### *11.3.1 Genomics*

The genome of an organism encompasses its entire hereditary information, encoded in DNA (or RNA for certain viruses). It comprises both genes and non-coding DNA sequences. Specifically, an organism's genome represents the complete DNA sequence of one set of chromosomes, such as one of the two sets found in each somatic cell of a diploid individual (Debnath et al., 2010). This genetic blueprint includes the instructions necessary for the organism to construct and sustain itself.

Genomics, as a branch of molecular biology, focuses on studying an organism's genome or genetic material. The emergence of genomics as a field within molecular biology has been driven by the application of innovative technologies aimed at unraveling the intricate biological functions encoded in the genome. Genomics, a foundational discipline within the realm of omics-based technologies, is the study of the complete set of genes, known as the genome, within an organism. This field has undergone transformative advancements with the advent of high-throughput sequencing technologies, enabling the rapid and cost-effective decoding of entire genomes (Venter et al., 2001). Genomics provides a comprehensive understanding of an organism's genetic makeup, allowing researchers to elucidate the structure, function, and regulation of genes. The information obtained from genomic studies is invaluable for diverse applications, including microbial bioprospecting.

In the context of microbial bioprospecting, genomics plays a pivotal role in identifying genes associated with desirable traits, such as the production of bioactive compounds, enzymes, and other industrially relevant molecules. Comparative genomics, which involves the comparison of genomes from different microbial strains, facilitates the identification of unique features and potential functions (Medini et al., 2005). This comparative approach guides the selection of promising microbial candidates for further exploration and exploitation. Moreover, genomics contributes to the understanding of microbial evolution, adaptation, and diversity. It enables the reconstruction of phylogenetic relationships among microorganisms, shedding light on the evolutionary processes that have shaped microbial communities (Woese et al., 1990). The identification of core and accessory genomes allows researchers to discern conserved functions and adaptive traits, providing insights into the ecological roles of microorganisms in various environments.

### *11.3.2 Metagenomics*

Metagenomics is a field of genomics that involves the direct analysis of DNA extracted from environmental samples, providing a powerful means to study microbial communities without the need for individual microbial isolates (Handelsman et al., 1998). This approach has revolutionized our understanding of microbial diversity, functional potential, and ecological roles in various environments, ranging from oceans and soils to the human microbiome. Key metagenomic techniques include shotgun metagenomics and 16S rRNA gene sequencing. Shotgun metagenomics involves the sequencing of all genomic DNA within a sample, providing insights into the entire genetic content of the microbial community. This approach allows for the discovery of novel genes, functional pathways, and potential interactions within complex microbial ecosystems. 16S rRNA gene sequencing, on the other hand, targets a specific phylogenetic marker gene to characterize the microbial diversity and community structure. This method is particularly useful for taxonomic profiling and identifying key microbial players in a given environment (Gilbert & Dupont, 2011).

Metagenomics has been applied in various fields, including environmental microbiology, biotechnology, and medicine. In environmental microbiology, metagenomics has unveiled the microbial diversity in extreme environments, elucidated nutrient cycling processes, and identified novel enzymes with industrial applications (Qin et al., 2010). In biotechnology, metagenomics has been instrumental in the discovery of enzymes for biofuel production, bioremediation, and the synthesis of valuable chemicals. In medicine, metagenomics has provided insights into the human microbiome, linking microbial community composition to health and disease states. The continuous advancements in high-throughput sequencing technologies, coupled with bioinformatics tools, have propelled metagenomics into a forefront position for studying microbial communities. Integrative analyses, such as functional metagenomics, allow for the functional characterization of microbial communities, shedding light on the roles microorganisms play in various ecosystems.

### *11.3.3 Transcriptomics*

Transcriptomics is a branch of molecular biology that involves the comprehensive study of RNA molecules within a cell or organism, providing insights into gene expression patterns and regulation. Transcriptomics is the field dedicated to studying transcriptomes, encompassing all RNA transcripts, both coding and non-coding, generated by the genome during development or in specific conditions (Sheng et al., 2017). Essentially, a transcriptome offers a snapshot of all transcripts present in a cell at a particular moment, providing insights into actively expressed genes (Lowe et al., 2017). Various techniques are employed in transcriptomics, including Serial/Cap Analysis of Gene Expression (SAGE/CAGE), Expressed Sequence Tag (EST), Suppression Subtractive Hybridization (SSH), microarrays, and RNA sequencing (RNA-Seq). These methods quantify gene expression in distinct cell populations affected by different treatments, diseases, or environmental factors at varying time points. SSH, a Polymerase Chain Reaction (PCR-based approach), amplifies differentially expressed cDNAs (complementary DNAs) fragments, and despite the advent of newer techniques, it remains widely used for gene expression profiling (Rusaini, 2018).

Microarray analysis, although less commonly used in recent years due to the advent of RNA-seq, remains a valuable method for examining gene expression on a genome-wide scale. DNA microarrays consist of microscopic DNA spots arranged on a solid substrate (Romanov et al., 2014). It involves the hybridization of labeled RNA samples to a microarray chip containing probes for thousands of genes. This technique allows for the simultaneous measurement of the expression levels of numerous genes, enabling the identification of transcriptional changes associated with specific biological conditions. As microarrays can only detect sequences homologous to those on the array, prior knowledge of the organism's genome, such as annotated genome sequences and ESTs, is necessary to create the array probes (Lowe et al., 2017). The advancement of microarray technologies and other high-throughput strategies is increasingly crucial in biology as it enables the simultaneous monitoring of expression levels for thousands of genes in a single experiment. To decipher the biological significance of these genetic profiles, it is essential to analyze the data in the context of the corresponding proteins they encode. This involves retrieving information regarding their biological, biochemical, or molecular functions. Consequently, there is a growing need for the development of computational tools that can compare and analyze these expression profiles in a manner suitable for biological interpretation (Subramanian et al., 2005).

RNA-Seq stands out as a revolutionary tool that utilizes recently developed deep-sequencing technologies along with computational methods for both mapping and quantifying transcriptomes (Lowe et al., 2017). RNA-seq has emerged as a powerful tool in transcriptomics, allowing for high-throughput sequencing of RNA molecules. This technique provides a comprehensive view of the transcriptome by quantifying gene expression levels, identifying alternative splicing events, and uncovering novel transcripts. The data generated from RNA-seq experiments can be analyzed to discern dynamic changes in gene expression under different conditions, providing valuable insights into cellular responses to stimuli, developmental processes, and disease mechanisms.RNA-Seq methodologies offer numerous advantages, including higher throughput, sensitivity, accuracy, long read lengths, and the absence of a requirement for prior knowledge of the organism's genome. These attributes have led to the rapid adoption of RNA-Seq over hybridization-based approaches, establishing it as the dominant transcriptomics technique over the last two decades (Su et al., 2014).

Quantitative reverse transcription PCR (qRT-PCR) is a highly sensitive and widely used method for validating and quantifying gene expression data obtained from high-throughput techniques. It provides a targeted approach, allowing researchers to focus on specific genes of interest and validate expression changes with high precision.This field plays a crucial role in understanding how the genetic information encoded in DNA is transcribed into functional RNA molecules, ultimately influencing the synthesis of proteins and the regulation of various cellular processes (Mauger & Deleuze, 2019).Transcriptomics has diverse applications across biological research, including the study of developmental processes, response to environmental stimuli, identification of biomarkers for diseases, and the exploration of gene regulatory networks. By unraveling the intricacies of gene expression, transcriptomics contributes to our understanding of the molecular basis of cellular functions and provides valuable information for applications in medicine, agriculture, and biotechnology.

## 11.3.4 Proteomics

Proteomics is a field of study focused on the large-scale analysis of proteins within a biological system, providing a comprehensive understanding of their structure, function, and interactions. The term "proteome" refers to the entire set of proteins expressed by a cell, tissue, or organism at a specific time under defined conditions. Proteomics aims to elucidate how proteins contribute to the complex network of biological processes and responses (Yates et al., 2009). Since proteins play a crucial role in carrying out most biological functions, it is essential to accurately assess changes in the proteome when cells undergo transitions in their states. Proteomic analyses involve the identification, quantification, and characterization of proteins, often leveraging mass spectrometry (MS) as a key tool. Liquid chromatography coupled with mass spectrometry (LC-MS/MS) is a widely used approach that allows for the high-throughput identification of proteins in complex samples. Other techniques, such as two-dimensional gel electrophoresis (2D-GE), protein microarrays, and gel-free methods, also contribute to the diverse toolkit of proteomic technologies (Paik et al., 2012).

Proteomic technologies rely on the capacity to separate intricate mixtures, making individual proteins more manageable for further processing using various techniques. Traditional methods, such as low-throughput sequencing through Edman degradation, are well-known. In contrast, more advanced proteomic techniques, particularly those with higher throughput, are based on mass spectrometry. Common approaches include peptide mass fingerprinting using simpler instruments and De novo repeat detection sequencing on instruments capable of multiple rounds of mass spectrometry. Recent successes highlight the critical role of mass spectrometry-based proteomics as an indispensable tool in molecular and cellular biology, as well as the emerging field of systems biology. These achievements encompass the investigation of protein-protein interactions on both a small and proteome-wide scale through affinity-based isolations, mapping numerous organelles, simultaneous exploration of the malaria parasite genome and proteome, and the generation of quantitative protein profiles from diverse species.

The capability of to not only identify but increasingly quantify thousands of proteins from complex samples is anticipated to have a significant impact on the fields of biology and medicine (Aebersold & Mann, 2003). Protein correlation profiling allows the analysis of multiprotein complexes that can be enriched by fractionation but are not purified to homogeneity. An example of this is the work of Andersen et al. (2003), who conducted a mass-spectrometry-based proteomic analysis of human centrosomes during the interphase of the cell cycle. This involved quantitatively profiling hundreds of proteins across several centrifugation fractions. The comprehensive identification of all protein-protein interactions within an organism represents a key goal in proteomics. A related objective is determining which protein pairs are present in the same protein complex. High-throughput techniques like yeast two-hybrid (Y2H) and affinity purification coupled with mass spectrometry (APMS) have been employed for detecting interacting proteins on a genomic scale. However, both Y2H and APMS methods exhibit notable false-positive rates. Beyond these high-throughput screens, characteristics of gene- or protein-pairs, aside from direct interaction detection, can offer valuable information about physical interactions. Therefore, there is a need to integrate multiple datasets and leverage their distinct predictive values to enhance the accuracy of predicting co-complexed relationships. Many proteins play roles in cellular processes as members of protein complexes of various sizes. It is believed that the combinatorial interactions among proteins form a crucial basis for the biological complexity observed in higher organisms (Zhang et al., 2004). Consequently, advancing our understanding of protein function is greatly facilitated by increased knowledge about protein-protein interactions and protein complexes.

The application of proteomics spans various biological disciplines, including molecular biology, medicine, and biotechnology. In molecular biology, proteomics aids in deciphering the intricate signaling pathways, protein-protein interactions, and post-translational modifications that regulate cellular functions. In medicine, proteomics plays a crucial role in biomarker discovery for disease diagnosis, prognosis, and therapeutic monitoring. Additionally, proteomics contributes to drug development by identifying potential drug targets and elucidating drug-protein interactions. In biotechnology, proteomics facilitates the optimization of protein expression in engineered organisms for industrial processes.

### *11.3.5 Metabolomics*

Metabolomics is a rapidly evolving field of study that involves the comprehensive analysis and profiling of small molecules, or metabolites, within a biological system. As one of the most recent additions to the "omics" fields, metabolomics has experienced rapid growth over the past decade. These metabolites include a diverse range of compounds such as amino acids, lipids, sugars, and other intermediates involved in cellular processes (Nicholson et al., 1999). Metabolites, being the end products of gene expression and cellular activities, exhibit high sensitivity to environmental changes. Consequently, metabolomics provides a real-time snapshot of the ongoing processes within an organism at a specific moment (Nguyen et al., 2018).

The analytical techniques employed in metabolomics, such as nuclear magnetic resonance (NMR) spectroscopy and mass spectrometry (MS), enable the identification and quantification of metabolites present in a given sample (Fiehn, 2002). These technologies are complemented by various data analysis approaches, including statistical methods and pathway mapping, to interpret the complex datasets generated in metabolomic studies. The application of metabolomics aims to characterize intrinsic metabolic alterations in biological samples under varying environmental conditions and identify biomarkers associated with these processes. Analytical platforms employed in metabolomics include Infrared Spectroscopy (IR), Raman Spectroscopy, Nuclear Magnetic Resonance (NMR), and various Mass Spectrometry (MS) techniques. The latter encompasses Direct

Infusion Mass Spectrometry (DI-MS), Matrix-Assisted Laser Desorption/Ionization Mass Spectrometry (MALDI-IMS), Capillary Electrophoresis-Mass Spectrometry (CE-MS), Gas Chromatography-Mass Spectrometry (GC-MS), and Liquid Chromatography-Mass Spectrometry (LC-MS). Among these, NMR and MS, owing to their high throughput and resolution capabilities, are the most widely utilized analytical tools (Nguyen et al., 2019).

Metabolomics finds applications across diverse scientific domains, including medicine, environmental science, agriculture, and systems biology. In medicine, metabolomics contributes to biomarker discovery for disease diagnosis, monitoring treatment responses, and understanding metabolic alterations associated with various disorders. In environmental science, metabolomics helps assess the impact of pollutants on organisms and ecosystems. In agriculture, metabolomics aids in optimizing crop yield and quality by studying plant metabolism under different conditions (Wishart, 2019). In systems biology, metabolomics data are integrated with genomics, transcriptomics, and proteomics to provide a holistic understanding of cellular functions. One of the key challenges in metabolomics is the sheer complexity of the metabolome, which is influenced by genetic, environmental, and lifestyle factors. Nonetheless, the field continues to advance, driven by innovations in analytical technologies, data analysis methods, and collaborative efforts to standardize experimental protocols and databases.

## 11.4 Applications of Omics-Based Technologies in Microbial Bioprospecting

Omics-based technologies play a pivotal role in microbial bioprospecting, offering powerful tools for the discovery, characterization, and optimization of microbial resources. The identification of molecular components involved in biological processes has deepened our comprehension of biochemical and metabolic pathways. This understanding has paved the way for the enhancement of organisms, either directly or indirectly, to maximize biomedical benefits. The progress observed in medical, pharmaceutical, chemical, nutrition, and agricultural industries are indebted to omics-based approaches (Harris et al., 2015). Initially, the vast amount of data generated by individual omic approaches was analyzed separately. However, recent research studies have integrated information from various tools, utilizing it to develop industrially viable organisms (Meng et al., 2014). Complementing high-throughput experimental methods, innovations in computational modeling and simulation have played a crucial role in unraveling the structural and functional foundations of biological processes (Segata et al., 2013; Kholodenko et al., 2012). The applications of omic based technologies in microbial bioprospecting are discussed below:

### 11.4.1 Next Generation Sequencing

In bioprospecting, the exploration of genes responsible for encoding enzymes involved in biosynthetic pathways of specific molecules is a fascinating area of study. In the past, prior to the advent of nucleic acid sequencing technologies, conventional research concentrated on identifying and characterizing the roles of one to a few genes. This conventional approach typically relied on traditional polymerase chain reaction (PCR) amplification of the target gene and subsequent sequencing using first-generation techniques, leaving the remainder of the genome unexplored. However, with the current affordability of DNA sequencing technologies, it has become feasible to obtain complete genomes for numerous strains. Ongoing studies indicate promising possibilities for sequencing the genomes of thousands of microorganisms in the coming decades. Next-generation sequencing (NGS), also commonly known as high-throughput sequencing, encompasses a set of techniques designed for the rapid reading of millions of nucleotides (Marx, 2013). This methodology involves sequencing extensive data using specific fragment libraries that exhibit enhanced efficiency compared to traditional sequencing techniques. Various genome-sequencing platforms have been developed for this purpose, including the Roche 454 sequencer, Illumina sequencer, and SOLiD sequencerwhich have made milestonecontributions to the omics field. The initial NGS system, Roche 454 FLX sequencer, was introduced by Roche (Mardis, 2008; Bentley et al., 2008; Valouev et al., 2008; Margulies et al., 2005).

Advancements in NGS technology have significantly improved the effectiveness of gene ontology by facilitating the identification of molecular interactions within a microbiome through annotations (Mills et al., 2013). NGS provides a more robust platform for cultivating a deeper understanding of the microorganisms of interest under established conditions (Rückert et al., 2015). The majority of current next-generation sequencing (NGS) technologies operate based on the sequencing by synthesis approach, categorizing second- and third-generation sequencing methods into short-read and real-time long-read technologies, respectively. Despite the diversity in NGS methodologies, the workflow typically involves three main steps: sample preparation, nucleic acid sequencing, and data analysis (Kumar et al., 2019).

The existence of reference genomes greatly facilitates the identification of novel strains through global and local alignment, as well as structural and functional annotation of the genomes. This streamlined process enables the identification of desirable strains for bioprospecting. Furthermore, the continuous progress in bioinformatics databases and analysis platforms has created a need for the development of more user-friendly tools and the training of additional experts in the field of bioinformatics. Next-generation sequencing (NGS) offers the ability to sequence millions of DNA reads in a single assay at a significantly reduced cost. This advantage has led to the widespread use of NGS methods across various applications, including variant identification through whole genome/exome resequencing, transcriptome profiling of tissues, microbial profiling, and the detection of genetic biomarkers for disease prognosis (Schuster, 2008; Suravajhala et al., 2016). By initially sequencing samples and applying bioinformatics techniques, NGS allows for the identification of both coding and noncoding regions. Furthermore, it enables the observation of protein-protein interactions and protein-ligand interactions. The confirmation of site-directed mutagenesis, a crucial step in many procedures aimed at improving the overall outcomes of industrial processes, can be verified, optimized, and validated using the same NGS-based technique (Babar et al., 2018). NGS is not only valuable for mutation discovery but also for metagenomic characterization in the research and development (R&D) sector of the industry. Although NGS is a relatively new technique, its integration into mainstream industrial biotechnology has the potential to enhance product quality and bioprocess yield significantly.

### *11.4.2 Cell Line Development and Cell Engineering*

Over the years, there have been significant advancements in the development of cell lines, driven by the need to reduce the time and costs associated with using complex hosts as production platforms. Recent research has demonstrated the potential of genomic and proteomic techniques in understanding intracellular changes when environmental conditions are altered, such as variations in pH, osmolarity, and temperature. Studies have identified expression changes in genes and proteins directly or indirectly associated with metabolic processes in various cell clones and lines (Bailey et al., 2002). While direct cell engineering has traditionally focused on manipulating single genes crucial in key metabolic and regulatory pathways to enhance cell lines, a recent shift has occurred towards adopting an indirect cell engineering approach. This approach utilizes genomic and proteomic techniques and tools to facilitate the discovery of novel targets for metabolic manipulation within the host cell line (Walsh, 2006). Two recently introduced technologies now offer a comprehensive high-throughput evaluation of changes occurring at both the gene (genomics) and protein (proteomics) levels, providing insights into productivity variations under different environmental conditions. These technologies enable the establishment of functional connections between cellular organization and productivity. Integrating genomics and proteomic analyses with cell engineering strategies brings us closer to comprehending cellular behavior during bioprocessing. Unraveling the biological pathways regulated by diverse environmental conditions contributes to a deeper understanding of the mechanisms governing cell survival and productivity. Delving into the role of these pathways may reveal potential targets for intervention, paving the way for the development of more robust and productive cell lines.

The most common proteomic technique, Two-dimensional polyacrylamide gel electrophoresis (2D-PAGE or 2DE), has seen significant advancements with the introduction of 2D-differential gel electrophoresis (DIGE) and improvements in quantitative analysis, image warping, and spot detection algorithms (Lilley &

Friedman, 2004; Miller et al., 2006). In addition to 2D-gel techniques, mass spectrometry (MS) has emerged as a powerful and alternative tool in proteomics, allowing the analysis of numerous proteins in complex biological samples through methods like MudPIT (multi-dimensional protein identification technology), SILAC (stable isotope labeling by amino acids), ICAT (isotope-coded affinity tags), iTRAQ, or 16O-to-18O for functional and quantitative proteomics (Ong & Mann, 2005; Roe & Griffin, 2006).In genomics, high-density oligonucleotide GeneChip microarrays are commonly employed for transcriptomics analysis to analyze gene expression. These technologies offer deeper insights into cell metabolism and physiology under process and manufacturing conditions (Wurm, 2004). These approaches have led to various current and potential improvements in cell lines of biopharmaceutical significance. The advent of genomic and proteomic tools has made indirect cell engineering a valuable strategic approach for enhancing biopharmaceutical cell lines. When combined with advancements in expression engineering, selection, and media development, it is anticipated to reduce development time and significantly increase cell line productivity, ultimately reducing overall costs for the next generation of approved products.

Genomic approaches have played a crucial role in the creation of cell lines with heightened resistance to signals that impede their expression capacity. Introduction of cyclins, protooncogenes, and certain growth factor genes has been employed to enhance production capabilities. However, it is noteworthy that modifications aimed at increasing production capacity may sometimes be associated with a decline in product quality. To address this, recombinant cell lines are now designed and developed with a focus on highly regulated posttranslational protein modification assemblies. These cell lines are engineered to respond effectively to specific controlling factors such as temperature, gaseous content, and/or certain chemical agents (Lai et al., 2013). These factors serve as inducing or controlling agents for the expression process, with different industries creating customized platforms to select the most desirable features. In the case of developing a recombinant cell line, a crucial initial step involves expressing the gene of interest in a transient expression system (Kim et al., 2012). This allows for the assessment of the gene's potential to be expressed under varying conditions. Once it is confirmed that the gene can successfully incorporate into a cell and be expressed, the gene is introduced into separately selected cells (Ram et al., 2016). Integration of the gene into the host-cell genome occurs at a random location, and specific clones with desirable characteristics are then chosen using appropriate screening techniques (Young, 2013).

With the emergence of genomic and proteomic tools, the processing of cell lines, including both their development and engineering, has experienced rapid advancements over the past few decades. Concurrently, proteomics techniques have been employed to explore the structural aspects of proteins, their folding processes, and posttranslational modifications. Furthermore, the tools of metabolomics and fluxomics offer valuable insights into unraveling the intricate metabolic machinery of cell systems. Moreover, the application of functional genomics and proteomics tools can be directly integrated into cell and metabolic engineering processes, facilitating the identification of novel targets for enhanced phenotypes. This approach not only provides immediate benefits but also contributes to acquiring fundamental knowledge about the genetic mechanisms involved in recombinant protein production.

### *11.4.3 Systems Biology for Holistic Understanding*

Integrating data from genomics, transcriptomics, proteomics, and metabolomics, systems biology provides a holistic understanding of microbial systems. This comprehensive approach allows for the modeling and optimization of microbial processes, enhancing bioprospecting efforts.

In the realm of industrial biotechnology, these methodologies are applied to devise effective strategies for manufacturing biological products. Many industrial biotechnology processes primarily aim to generate products from sustainable resources utilizing microorganisms. To achieve optimal efficiency in terms of both output and cost, it is imperative to develop "hyper-productive" microbial strains (Costanza et al., 2012). While traditional genetics played a role in reaching these objectives through labor-intensive cycles of mutation screening, the emergence of metabolic engineering, grounded in systems biology tools, aims to enhance the production of dependable and highly efficient microbial strains. This approach involves manipulating the chemical and metabolic networks within the biological system. Synthetic biology and metabolic engineering utilize mathematical modeling of biological networks to identify the optimal ratios of reactants that lead to the production of specific products. They also pinpoint the critical points in pathways that can be exploited to increase process yields (Wang & Wu, 2015).

The fundamental objective of synthetic biology is to enhance and standardize biological systems, thereby increasing control over them. The progress in genomics, coupled with information systems, has played a pivotal role in unraveling the intricacies of biological processes. The advancement of computer-aided design now allows for the convenient design of genes and proteins. Additionally, efficient sequencing methods facilitate the optimization and screening of engineered biological organisms. Beyond genetic modifications, proteins can undergo alterations to enhance their efficiency. This includes the design of intracellularly produced fusion proteins with increased stability and the creation of conjugated proteins composed of various combinations of proteins and/or peptides, which contribute to improving system activity and efficiency. Once a protein molecule is designed, computational modeling can be employed to explore various structural features before actual modification or synthesis. These modeling techniques are also valuable for studying the interactions among different biological components and small ligands/inhibitors. Molecular simulation techniques play a crucial role in studying essential biomolecular interactions during processes such as replication, transcription, translation, and regulation. Similarly, fluxomics approaches prove useful in examining enzyme kinetics and the rates of metabolic processes by physically measuring reaction outputs and rates to predict biological mechanisms (Babar et al., 2018).

The tools of systems biology play a crucial role in enabling the strategic design of both the bioprocess and the resulting product. These tools prove beneficial across various stages, assisting not only in the production process but also in both upstream and downstream bioprocessing activities. The comprehensive and interdisciplinary approach of systems biology contributes to the development of optimal biopharmaceuticals. By supporting the entire lifecycle, from the conceptualization and synthesis stages to the ultimate utilization of biopharmaceuticals, systems biology enhances the efficiency and effectiveness of the overall production process.

## 11.5 Challenges in Applications of Omic Technologies

Throughout human history, the rapid evolution of technologies like high-throughput sequencing and mass spectrometry has given rise to the concept of "omics" and has significantly advanced the methodologies for systematically examining cellular systems. The overarching aim of "omic" approaches is to gain a comprehensive and integrated understanding of biology. This involves studying all biological processes to identify various components, such as genes, RNA, proteins, and metabolites, collectively rather than focusing on each component individually. Despite the abundance and diversity of both experimental and computational approaches available for omics exploration, the translation of knowledge obtained from delving into various omics levels into practical applications in clinical settings is still in its early stages of development.

Their primary concerns revolve around issues of reproducibility and specificity, both of which are closely interconnected and interdependent. The generation of a substantial volume of data coupled with the selective interpretation of results to align with proposed hypotheses contributes to these challenges. Inappropriate choices in the selection of interpretation and validation tools further exacerbate the issues, impacting the reliability of the entire process. In the context of devising and optimizing strategies for industrial processes, unreliable procedures are generally avoided (Gomez-Cabrero et al., 2014). Furthermore, when optimizing strategies for the genetic manipulation of commercially viable organisms, it is essential to account for the role of various confounding variables. Many experimental techniques yield disparate results when studied under different environmental and experimental conditions. Consequently, the sensitivity of these methods must be considered when examining the overall dynamics of a process. To ensure robust findings, the outcomes of any biological process should only be presented after thorough verification and validation under various experimental conditions (Machado et al., 2015). Technical challenges encompass various aspects, such as significantly decreasing sample amounts without compromising statistical accuracy, especially when dealing with rare and precious samples. Additionally, finding ways to reduce replicates without adversely impacting output stability and enhancing the accuracy of the output by minimizing both false negative (FN) and false positive (FP) rates pose considerable issues. Comparatively, the reproducibility of MS-based omics tends to be lower than sequencing-based omics. Moreover, impediments to the knowledge acquisition process include the heterogeneity observed across studies in terms of study design, sample size, sampling approach, treatment methods, and follow-up duration. This heterogeneity results in limited generalizability of the diverse outputs consistently reported in succession (Dai & Shen, 2022).

The rapid progress in 'omics technology presents several challenges in making them applicable for hazard assessment, particularly concerning regulatory submissions (Sauer et al., 2017). Handling sets of extensive 'big data' requires the application of intricate technologies and methodologies, along with specific expertise, to extract information relevant to hazard assessment. The knowledge in this domain is part of a swiftly evolving scientific field, which many scientists in regulatory settings may not be fully acquainted with. Consequently, the limited adoption of 'omics technologies in regulatory practices is not only tied to the absence of best practices frameworks and quality criteria but also to a lack of confidence in the analysis and uncertainty about what constitutes sufficient data (Healy et al., 2016). Beyond technical concerns, addressing regulatory and legal aspects is crucial to establish the regulatory applicability of 'omics data, and these issues are more intricate than the standardization and validation challenges. Despite legal complexities, it is imperative to establish robust, transparent processes that prevent the compromise of 'omics study results due to uncertainties arising during data generation, storage, curation, and processing. Such measures are essential to instill confidence in the regulatory use of these technologies (McConnell et al., 2014).

Further, the gap between academia and industry often stands as a fundamental reason for the limited adoption of omics approaches in the biotechnology industry. The criteria for success in genetic modification within academic settings can significantly differ from those in an industrial setup, where a high-commercial outcome typically takes precedence. Furthermore, financial constraints often lead many industrial units to allocate insufficient funds for establishing appropriate omics facilities. Addressing these challenges is crucial when formulating a comprehensive "omics-for-industry" policy (Babar et al., 2018).

## 11.6 Future Perspectives

As we look ahead to the future of omics-based technologies in microbial bioprospecting, several promising directions emerge, reflecting a dynamic landscape of innovation and discovery. Firstly, the integration of multi-omics data will continue to be a pivotal focus, allowing for a more comprehensive understanding of microbial systems by combining genomics, transcriptomics, proteomics, and metabolomics datasets (Ballereau et al., 2013). This holistic approach will not only enhance our ability to unravel complex microbial interactions but will also facilitate the engineering of microbes with tailored functionalities for specific applications. Advancements in single-cell omics technologies will likely play a transformative role, enabling the study of individual microbial cells and providing unparalleled insights into the heterogeneity within microbial communities and their responses to environmental stimuli. This level of granularity is particularly valuable when working with complex microbial ecosystems, allowing researchers to uncover hidden functional traits (Head et al., 2014).

Functional metagenomics, which explores the genetic potential of uncultivable microbes directly from environmental samples, is poised for further development. This approach facilitates the identification of novel genes and functional traits, offering a promising avenue for bioprospecting in diverse and challenging environments. The field of synthetic biology is expected to advance, allowing for the design and construction of customized microbial pathways geared towards the production of specific compounds. This engineering capability presents significant potential for tailoring microbial strains to fulfill specific industrial or biomedical requirements, thereby broadening the scope of microbial bioprospecting applications. Moreover, in the context of optimizing microbial manufacturing processes to enhance yield and reduce costs, it is imperative to consider factors such as the selection and design of fermentation reactors, substrate pretreatment, wastewater treatment, and downstream separation and purification. Addressing these aspects ensures the comprehensive improvement of microbial manufacturing performance, ultimately benefiting humanity (Wan et al., 2023). The growing prominence of environmental DNA (eDNA) sampling techniques will facilitate non-invasive exploration of microbial diversity in various ecosystems. This approach will be crucial for understanding microbial communities in their natural habitats and identifying potential candidates for bioprospecting.

In the realm of omics-based studies, machine learning and advanced data analytics are becoming increasingly integral. These technologies play a crucial role in enhancing our capacity to interpret extensive datasets, pinpoint key genetic targets, foresee microbial behavior, and refine bioprospecting strategies. The incorporation of computational diagnostic methods, particularly through machine and deep learning algorithms, offers advantages such as greater consistency, impartiality, and a reduced susceptibility to human errors. Employing machine learning algorithms as diagnostic modeling tools presents several benefits. However, it is essential to ensure the proper training of these models for a comprehensive understanding of biological systems and gaining valuable insights into microbial bioprospecting (Prasad et al., 2021). As the field advances, there will be an increasing need for ethical considerations and regulatory frameworks to govern the collection, use, and commercialization of microbial resources. Striking a balance between scientific exploration and responsible bioprospecting practices will be imperative to ensure the sustainable and ethical utilization of microbial diversity (McConnell et al., 2014). The future of omics-based technologies in microbial bioprospecting promises to be dynamic and transformative, with ongoing advancements poised to unlock new dimensions of microbial diversity for diverse applications. These future directions underscore the importance of interdisciplinary collaboration, technological innovation, and ethical stewardship in shaping the trajectory of microbial bioprospecting research.

## 11.7 Conclusion

The exploration of microbial bioprospecting has been significantly transformed by the advent and integration of omics-based technologies. This chapter has delved into the diverse applications of genomics, metagenomics, transcriptomics, proteomics, metabolomics, functional genomics, and systems biology in the pursuit of uncovering and optimizing microbial resources. These advanced tools have revolutionized our ability to decipher the genomic landscape of microorganisms, analyze their functional gene expression profiles, and understand their metabolic capabilities on a systems level. The genomic bioprospecting approach, guided by comparative genomics, has empowered researchers to identify genes associated with valuable traits and select microbial candidates with specific functionalities. Functional genomics has facilitated the systematic analysis of gene functions, guiding the engineering of microbial pathways for enhanced bioprospecting outcomes. The holistic perspective offered by systems biology, integrating data from multiple omics approaches, has provided a comprehensive understanding of microbial systems. This has not only enhanced our fundamental knowledge but has also paved the way for the rational design and optimization of microbial strains for diverse biotechnological applications.

As we stand at the forefront of microbial bioprospecting, omics-based technologies serve as indispensable tools, driving innovation and accelerating the discovery of valuable microbial resources. The integration of these advanced techniques into biotechnological and industrial applications holds great promise for addressing global challenges in pharmaceuticals, agriculture, and renewable energy. Looking ahead, the continued evolution of omics technologies will undoubtedly shape the future of microbial bioprospecting, unlocking new frontiers in biotechnology and sustainable resource utilization.

## References

Aebersold, R., & Mann, M. (2003). Mass spectrometry-based proteomics. *Nature, 422*, 198–207.

Amer, B., & Baidoo, E. E. K. (2021). Omics-driven biotechnology for industrial applications. *Frontiers in Bioengineering and Biotechnology, 9*, 19.

Andersen, J. S., Wilkinson, C. J., Mayor, T., Mortensen, P., Nigg, E. A., & Mann, M. (2003). Proteomic characterization of the human centrosome by protein correlation profiling. *Nature, 426*, 570–574.

Babar, M. M., Afzaal, H., Pothineni, V. R., Zaidi, N.-U. S., Ali, Z., Zahid, M. A., & Gul, A. (2018). Omics approaches in industrial biotechnology and bioprocess engineering. In *Omics technologies and bio-engineering: Towards improving quality of life* (Vol. 2, pp. 251–269). Elsevier.

Bailey, J. E., Sburlati, A., Hatzimanikatis, V., Lee, K., Renner, W. A., & Tsai, P. S. (2002). Inverse metabolic engineering: A strategy for directed genetic engineering of useful phenotypes. *Biotechnology and Bioengineering, 79*, 568–579.

Ballereau, S., Glaab, E., Kolodkin, A., Chaiboonchoe, A., Biryukov, M., Vlassis, N., et al. (2013). Functional genomics, proteomics, metabolomics and bioinformatics for systems biology. In *Systems biology* (pp. 3–41). Springer.

Bansal, M., Tiwari, N., Sharma, J. G., Verma, P., & Shah, M. P. (2022). Revolution in microbial bioprospecting via the development of omics-based technologies. In *Bioprospecting of microbial diversity* (pp. 27–46). Elsevier. https://doi.org/10.1016/B978-0-323-90958-7.00008-X

Bentley, D. R., Balasubramanian, S., Swerdlow, H. P., Smith, G. P., Milton, J., Brown, C. G., et al. (2008). Accurate whole human genome sequencing using reversible terminator chemistry. *Nature, 456*, 53–59. https://doi.org/10.1038/nature07517

Berger, B., Peng, J., & Singh, M. (2013). Computational solutions for omics data. *Nature Reviews Genetics, 14*(5), 333–346.

Costanza, J., Carapezza, G., Angione, C., Lio, P., & Nicosia, G. (2012). Robust design of microbial strains. *Bioinformatics, 28*(23), 3097–3104.

Dai, X., & Shen, L. (2022). Advances and trends in omics technology development. *Frontiers in Medicine, 9*. https://doi.org/10.3389/fmed.2022.911861

Debnath, M., Prasad, G. B. K. S., & Bisen, P. (2010). *Omics Technology*. https://doi.org/10.1007/978-90-481-3261-4

Fiehn, O. (2002). Metabolomics – The link between genotypes and phenotypes. *Plant Molecular Biology, 48*(1–2), 155–171.

Gilbert, J. A., & Dupont, C. L. (2011). Microbial metagenomics: Beyond the genome. *Annual Review of Marine Science, 3*, 347–371.

Golden, J. S., & Handfield, R. (2014). The emergent industrial bioeconomy. *Industrial Biotechnology, 10*(6), 371–375.

Gomez-Cabrero, D., Abugessaisa, I., Maier, D., Teschendorff, A., Merkenschlager, M., Gisel, A., et al. (2014). Data integration in the era of omics: Current and future challenges. *BMC Systems Biology, 8*(Suppl. 2), I1.

Handelsman, J. (2004). Metagenomics: Application of genomics to uncultured microorganisms. *Microbiology and Molecular Biology Reviews, 68*(4), 669–685.

Handelsman, J., Rondon, M. R., Brady, S. F., Clardy, J., & Goodman, R. M. (1998). Molecular biological access to the chemistry of unknown soil microbes: A new frontier for natural products. *Chemistry & Biology, 5*(10), R245–R249.

Harris, T., Papadopoulos, S., & Goldstein, D. B. (2015). Academic-industrial partnerships in drug discovery in the age of genomics. *Trends in Biotechnology, 33*(6), 320–322.

Head, S. R., Komori, H. K., LaMere, S. A., et al. (2014). Library construction for next-generation sequencing: Overviews and challenges. *BioTechniques, 56*, 61–77. https://doi.org/10.2144/000114133

Healy, M. J., Tong, W., Ostroff, S., Eichler, H. G., Patak, A., Neuspiel, M., et al. (2016). Regulatory bioinformatics for food and drug safety. *Regulatory Toxicology and Pharmacology, 80*, 342–347.

Keasling, J. D. (2012). Synthetic biology and the development of tools for metabolic engineering. *Metabolic Engineering, 14*(3), 189–195.

Kholodenko, B., Yaffe, M. B., & Kolch, W. (2012). Computational approaches for analyzing information flow in biological networks. *Science Signaling, 2002961.*

Kim, J. Y., Kim, Y.-G., & Lee, G. M. (2012). CHO cells in biotechnology for production of recombinant proteins: Current state and further potential. *Applied Microbiology and Biotechnology, 93*(3), 917–930.

Kumar, K. R., Cowley, M. J., & Davis, R. L. (2019). Next-generation sequencing and emerging technologies. *Seminars in Thrombosis and Hemostasis, 45*(07), 661–673. https://doi.org/10.1055/s-0039-1688446

Lai, T., Yang, Y., & Ng, S. K. (2013). Advances in mammalian cell line development technologies for recombinant protein production. *Pharmaceuticals, 6*(5), 579–603.

Lilley, K. S., & Friedman, D. B. (2004). All about DIGE: Quantification technology for differential-display 2D-gel proteomics. *Expert Review of Proteomics, 1*, 401–409.

Low, T. Y., van Heesch, S., van den Toorn, H., Giansanti, P., Cristobal, A., Toonen, P., et al. (2013). Quantitative and qualitative proteome characteristics extracted from in-depth integrated genomics and proteomics analysis. *Cell Reports, 5*(5), 1469–1478.

Lowe, R., Shirley, N., Bleackley, M., Dolan, S., & Shafee, T. (2017). Transcriptomics technologies. *PLoS Computational Biology, 13*, e1005457.

Machado, D., Zhuang, K. H., Sonnenschein, N., & Herrgard, M. J. (2015). Editorial: Current challenges in modeling cellular metabolism. *Frontiers in Bioengineering and Biotechnology, 3*, 193.

Mardis, E. R. (2008). Next-generation DNA sequencing methods. Annu. Rev. Genomics Hum. *Genet. 9*, 387–402.

Margulies, M., Egholm, M., Altman, W. E., Attiya, S., Bader, J. S., Bemben, L. A., et al. (2005). Genome sequencing in microfabricated high-density picolitre reactors. *Nature, 437*, 376–380. https://doi.org/10.1038/nature03959

Mauger, F., & Deleuze, J.-F. (2019). Technological advances in studying epigenetics biomarkers of prognostic potential for clinical research. In S. Sharma (Ed.), *Translational epigenetics, prognostic epigenetics* (Vol. 15, pp. 45–83). Academic. https://doi.org/10.1016/B978-0-12-814259-2.00003-0

Marx, V. (2013). Next-generation sequencing: the genome jigsaw. *Nature, 501*(7466), 263–268.

McConnell, E. R., Bell, S. M., Cote, I., Wang, R. L., Perkins, E. J., Garcia-Reyero, N., et al. (2014). Systematic Omics Analysis Review (SOAR) tool to support risk assessment. *PLoS One, 9*, e110379.

Medini, D., et al. (2005). The microbial pan-genome. *Current Opinion in Genetics & Development, 15*(6), 589–594.

Meng, C., Kuster, B., Culhane, A. C., & Gholami, A. M. (2014). A multivariate approach to the integration of multi-omics datasets. *BMC Bioinformatics, 15*(1), 1.

Miller, I., Crawford, J., & Gianazza, E. (2006). Protein stains for proteomic applications: Which, when, why? *Proteomics, 6*, 5385–5408.

Mills, J. D., Kawahara, Y., & Janitz, M. (2013). Strand-specific RNA-Seq provides greater resolution of transcriptome profiling. *Current Genomics, 14*, 173–181.

National Research Council (US) Committee on Metagenomics: Challenges and Functional Applications. Washington (DC): National Academies Press (US); 2007.ISBN-13: 978-0-309-10676-4.

Nguyen, T. V., Alfaro, A. C., Young, T., Ravi, S., & Merien, F. (2018). Metabolomics study of immune responses of New Zealand Greenshell™ mussels (Perna canaliculus) infected with pathogenic vibrio sp. *Marine Biotechnology, 20*, 396–409.

Nguyen, T. V., Alfaro, A. C., Young, T., & Merien, F. (2019). Tissue-specific immune responses to vibrio sp. infection in mussels (Perna canaliculus): A metabolomics approach. *Aquaculture, 500*, 118–125.

Nicholson, J. K., Lindon, J. C., & Holmes, E. (1999). 'Metabonomics': Understanding the metabolic responses of living systems to pathophysiological stimuli via multivariate statistical analysis of biological NMR spectroscopic data. *Xenobiotica, 29*(11), 1181–1189.

Ong, S. E., & Mann, M. (2005). Mass spectrometry-based proteomics turns quantitative. *Nature Chemical Biology, 1*, 252–262.

Prasad, A. et al. (2021). Next Generation Sequencing. In: Singh, V., Kumar, A. (eds) Advances in Bioinformatics. Springer, Singapore. https://doi.org/10.1007/978-981-33-6191-1_14

Paik, Y. K., Jeong, S. K., Omenn, G. S., Uhlen, M., Hanash, S., & Cho, S. Y. (2012). The chromosome-centric human proteome project for cataloging proteins encoded in the genome. *Nature Biotechnology, 30*(3), 221–223.

Qin, J., Li, R., Raes, J., Arumugam, M., Burgdorf, K. S., Manichanh, C., et al. (2010). A human gut microbial gene catalogue established by metagenomic sequencing. *Nature, 464*(7285), 59–65.

Ram, K., Hatton, D., Ahuja, S., Bender, J., Hunter, A., & Turner, R. (2016). *Protein production in eukaryotic cells* (pp. 1–39). Springer.

Roe, M. R., & Griffin, T. J. (2006). Gel-free mass spectrometry-based high throughput proteomics: Tools for studying biological response of proteins and proteomes. *Proteomics, 6*, 4678–4687.

Romanov, M., Ravi, K. V., & Chan, E. S. (2014). Temporally-correlated activity streams for conferences. *United States Patent US, 8*(812), 510.

Rückert, C., Albersmeier, A., Busche, T., Jaenicke, S., Winkler, A., Fri, Ó. H., et al. (2015). Complete genome sequence of Streptomyces lividans TK24. *Journal of Biotechnology, 199*, 21–22.

Rusaini, O. L. (2018). Suppression subtractive hybridization in penaeid prawns: An approach in identifying diseases and differentially expressed genes. *Journal of Aquatic Marine Biology, 7*, 00177.

Schuster, S. C. (2008). Next-generation sequencing transforms today's biology. *Nature Methods, 5*, 16–18.

Segata, N., Boernigen, D., Tickle, T. L., Morgan, X. C., Garrett, W. S., & Huttenhower, C. (2013). Computational meta'omics for microbial community studies. *Molecular Systems Biology, 9*(1), 666.

Sheng, J., Zheng, X., Wang, J., Zeng, X., Zhou, F., et al. (2017). Transcriptomics and proteomics reveal genetic and biological basis of superior biomass crop Miscanthus. *Scientific Reports, 7*, 1–3.

Su, L. T., Li, J. W., Liu, D. Q., Zhai, Y., Zhang, H. J., et al. (2014). A novel MYB transcription factor, GmMYBJ1, from soybean confers drought and cold tolerance in Arabidopsis thaliana. *Gene, 538*, 46–55.

Sauer, U. G., Deferme, L., Gribaldo, L., Hackermüller, J., Tralau, T., van Ravenzwaay, B., Yauk, C., Poole, A., Tong, W., & Gant, T. W. (2017). The challenge of the application of 'omics technologies in chemicals risk assessment: Background and outlook. *Regulatory Toxicology and Pharmacology, 91* (Suppl 1), S14–S26. https://doi.org/10.1016/j.yrtph.2017.09.020

Subramanian, A., Tamayo, P., Mootha, V. K., Mukherjee, S., Ebert, B. L., Gillette, M. A., Paulovich, A., Pomeroy, S. L., Golub, T. R., Lander, E. S., & Mesirov, J. P. (2005). Gene set enrichment analysis: A knowledge-based approach for interpreting genome-wide expression profiles. *Proceedings of the National Academy of Sciences, 102*, 15545–15550.

Suravajhala, P., Kogelman, L. J. A., & Kadarmideen, H. N. (2016). Multi-omic data integration and analysis using systems genomics approaches: Methods and applications in animal production, health and welfare. *Genetics, Selection, Evolution, 48*, 38. https://doi.org/10.1186/s12711-016-0217-x

Valouev, A., Ichikawa, J., Tonthat, T., Stuart, J., Ranade, S., Peckham, H., et al. (2008). A high-resolution, nucleosome position map of C. elegans reveals a lack of universal sequence-dictated positioning. *Genome Research, 18*, 1051–1063. https://doi.org/10.1101/gr.076463.108

Venter, J. C., et al. (2001). The sequence of the human genome. *Science, 291*(5507), 1304–1351.

Walsh, G. (2006). Biopharmaceutical benchmarks 2006. *Nature Biotechnology, 24*, 769–776.

Wan, S., Liu, X., Sun, W. et al. Current advances for omics-guided process optimization of microbial manufacturing. *Bioresour. Bioprocess. 10*, 30 (2023). https://doi.org/10.1186/s40643-023-00647-2

Wang, F.-S., & Wu, W.-H. (2015). Optimal design of growth-coupled production strains using nested hybrid differential evolution. *Journal of the Taiwan Institute of Chemical Engineers, 54*, 57–63.

Wishart, D. S. (2019). Metabolomics for investigating physiological and pathophysiological processes. *Physiological Reviews, 99*(4), 1819–1875.

Woese, C. R., et al. (1990). Towards a natural system of organisms: Proposal for the domains archaea, bacteria, and Eucarya. *Proceedings of the National Academy of Sciences, 87*(12), 4576–4579.

Wurm, F. M. (2004). Production of recombinant protein therapeutics in cultivated mammalian cells. *Nature Biotechnology, 22*, 1393–1398.

Xu, N., Ye, C., & Liu, L. (2018). Genome-scale biological models for industrial microbial systems. *Applied Microbiology and Biotechnology, 102*(8), 3439–3451.

Yates, J. R., Ruse, C. I., & Nakorchevsky, A. (2009). Proteomics by mass spectrometry: Approaches, advances, and applications. *Annual Review of Biomedical Engineering, 11*, 49–79.

Young, J. D. (2013). Metabolic flux rewiring in mammalian cell cultures. *Current Opinion in Biotechnology, 24*(6), 1108–1115.

Zhang, L. V., Wong, S. L., King, O. D., & Roth, F. P. (2004). Predicting co-complexed protein pairs using genomic and proteomic data integration. *BMC Bioinformatics, 5*, 38.

# Chapter 12
# Microbial Nanowires: Future of Bioenergy Applications

## Contents



## 12.1 Introduction

The depleting concentration of fossil fuels and the ever-increasing global demand of energy have created the necessity of the production of alternative fuels to replace the traditional sources in response to the increased greenhouse gas (GHG) deposition into the atmosphere, causing significant climatic changes. These changes could have catastrophic consequences in the near future, including rising temperatures and rising sea levels. The utilization of fossil fuels for generation of electricity and heat and for transportation accounts to about 25% and 14% of the total GHG emissions, respectively (IPCC, 2014). Therefore, the production of economically viable



P. Saini, P. Mishra, *Bioprospecting of Microbial Resources for Agriculture, Environment and Bio-chemical Industry*, https://doi.org/10.1007/978-3-031-63844-2_12

and environmentally friendly renewable energy fuels is the greatest need in the world, demonstrating the possibility of simultaneously replacing conventional fuels and reducing environmental concerns. The use of multifunctional microorganisms to produce renewable energy fuels from the biomass and bio- wastes could greatly reduce this looming concern. In recent years, there has been a growing interest in the use of microorganisms to produce various biofuels (Liao et al., 2016), especially due to the metabolic diversity of the different microorganisms that support the production of biofuels from various favorable substrates. For example, most bacteria can easily convert sugars into ethanol while cellulolytic microbes can utilize plant-driven substrates. Cyanobacteria and microalgae have the potential to reduce atmospheric CO2 into biofuels through photosynthesis whereas methanotrophs can use methane to produce methanol (Liao et al., 2016). Additionally, some of the bacteria such as *Geobacter sulfurreducens* and *Shewanella oneidensis* exhibit specific "molecular mechanisms" that facilitate the transfer of electrons from the outer microbial membrane to conductive surfaces (Kracke et al., 2015), subsequently, this functionality can be utilized in developing bioelectrochemical devices for biohydrogen and bioelectricity generation.

Microbial nanowires are electrically conductive filaments that facilitate long-range extracellular electron transfer. A model of electron transport along nanowires of *Shewanella oneidensis* lies between electron hopping/tunneling filaments adorned with cytochromes. Nanowires of *Geobacter sulfurreducens* consist of pili with metallic conductivity due to overlapping pi–pi orbitals of aromatic amino acids. Geobacter species nanowires participate in direct interspecies electron transfer (DIET), which may be an important cotrophicmode for the conversion of organic wastes to methane. Nanowire arrays give Geobacter biofilms electrical conductivity, which converts organic compounds to electricity in microbial fuel cells (MFCs) and increased nanowire production is the only genetic manipulation demonstrated to produce strains with improved current generation capabilities. The introduction of nanowires, or nanowire mimics, could enhance other bioenergy strategies that depend on extracellular electron exchange, like microbial electrosynthesis. The similarities between microbial nanowires and synthetic conductive polymers indicate potential energy-related applications. Nanowires produced by the oxyphototrophic cyanobacterium Synechocystis PCC6803 and the fermentative thermophilic *Pelotomaculum thermopropionicum* demonstrated that conductive appendages are not unique to dissimilatory metal reducing bacteria and in fact can represent an efficient strategy for electron transfer and energy distribution.

Protein nanowires can enhance the potential distance of electronic interaction between a cell and its environment, facilitating electron exchange with insoluble minerals and other cells. Enhanced expression of protein nanowires is a strategy to increase bimembrane conductance and current generation in bioelectrochemical devices (Lovley, 2017b). Protein nanowires produced by microbes or developed in vitro using microbe-inspired designs, are considered as "green," sustainable electronic materials with physical properties and functional potential for applications possible beyond with other nanowire materials (Lovley, 2017a, b; Creasey et al., 2018; Gutermann & Gazit, 2018; Ing et al., 2018; Ueki et al., 2019). A wide variety of microorganisms is able to produce protein nanowires (Reguera et al., 2005; Holmes et al., 2016; Sure et al., 2016; Tan et al., 2017; Walker et al., 2018, 2019), but in most cases, their structure and mechanism still need intensive investigation.

The process by which microbial nanowires are produced is believed to involve a number of different biological and biochemical mechanisms. One key step is the generation of energy through bacterial metabolism, which involves the oxidation of organic compounds and the transfer of electrons to an electron transport chain. This electron transport chain is thought to play a critical role in the production of microbial nanowires. As electrons move through the chain, they are transferred to specialized proteins that are involved in the assembly of the nanowires. These proteins include pilin-like proteins, cytochromes, and other electron transfer proteins that are thought to facilitate the formation of a conductive filament. Once these proteins have been synthesized, they are transported to the outer membrane of the bacterial cell where they are assembled into the nanowire. The assembly process is thought to be mediated by a variety of different factors, including protein-protein interactions, metal binding, and self-assembly. The resulting nanowires are long, thin filaments that can span significant distances, allowing for long-range electron transport. In addition to their role in electron transfer, microbial nanowires are also thought to play a role in biofilm formation. The extracellular matrix that surrounds bacterial cells contains a variety of polysaccharides, proteins, and other molecules that can interact with the nanowires to form a stable biofilm. This biofilm may help to protect the bacteria from environmental stresses and facilitate communication between cells. Overall, the mechanism of microbial nanowire production is a complex and multifaceted process that involves a number of different biological and biochemical factors. While much is still unknown about this process, ongoing research is helping to shed light on the properties and potential applications of these unique and fascinating structures. The chapter summarizes the recent literature on electron transport mechanisms, microbial nanowires and their potential role in bioenergy applications.

## 12.2 Flavin Electron Shuttles

An external electron exchange mechanism is the reduction of electron acceptors by microorganisms to generate electron carrying molecules that can serve as electron donors or electron acceptor electrodes for the microorganisms. Interspecific H2 transfer, in which electron donating organisms reduce protons to H2 whereas electron-accepting organisms oxidize H2 via reduction of the electron acceptor, is an important example of this form of electron exchange (Sieber et al., 2012). Organic electron shuttles, either produced by cells (eg. flavins, phenazines, cysteines) or commonly found in the extracellular environment (e.g. humus) can facilitate transfer of electrons to electrodes as well as interspecies electron transfer (Lovley, 2012; Liu et al., 2012; Brutinel & Gralnick, 2012). These electron shuttles, as well as terminal electron acceptors such as metals or electrodes, can be reduced by active redox molecules on the outer surface, such a c-type cytochromes (Kotloski & Gralnick, 2013; Voordeckers et al., 2010; Orellana et al., 2013; Inoue et al., 2010).

Conductive materials such as granular activated carbon (GAC) or magnetite also facilitates the exchange of electrons between species (Liu et al., 2012; Kato et al., 2012a, b). *Shewanella oneidensis* strain MR1 has been extensively studied for its ability to permeate a variety of soluble and insoluble electron acceptors. The ability to breathe insoluble substrates is defined as extracellular electron transfer and can occur by direct contact or electron shuttle in *S. oneidensis*. Flavin electron shuttle is the primary extracellular electron transfer mechanism used by *Shewanella oneidensis* bacteria.

All cultures of Shewanella are examined for the accumulation of riboflavin (B2) and flavin mononucleotide (FMN) in the supernatant. Both riboflavin and FMN act as electron shuttles to accelerate the reduction of insoluble substrates (Voordeckers et al., 2010; Orellana et al., 2013), including various forms of Fe(III) oxides (Inoue et al., 2010), and also facilitate the observation of redox gradients (Kato et al., 2012a). Secreted flavin is reduced by the Mtr airway in MR-1 (Kato et al., 2012b), and the crystal structure of the outer membrane-associated ten-heme cytochrome paralog MtrC reveals two solvent-exposed heme groups near FMN. The binding domain (Malvankar & Lovley, 2012), provided the maker with biochemical insight into how the flavin electron shuttle facilitates respiration. In the absence of outer membrane cytochromes, MR-1 is unable to shuttle electrons or directly access respiratory insoluble electron acceptors (Kato et al., 2012b; Gorby et al., 2006). Extracellular electron transfer by microbes is critical for the geochemical cycling of metals, bioremediation, and biocatalysis using electrodes. However, the contribution of electron shuttle versus direct contact to total extracellular electron transfer is unclear.

## 12.3 Electrically Conductive Pili

From early investigations (Gorby & Lovley, 1991; Lovley et al., 1993), to the present study on the mechanisms for extracellular electron exchange have focused primarily on cytochromes in Geobacter species (Lovley et al., 2004; Lovley, 2006; Shi et al., 2016). Mostly the researches focused on c-type cytochromes localized in the periplasm and outer membrane of the microorganisms. Figure 12.1 illustrates the bacterial electron transport processes.

The most well studied Geobacter protein nanowire is electrically conductive pili (e-pili). A significant number of studies have shown that the type IV pilin, PilA of *G. sulfurreducens*, assembles into e-pili (Reguera et al., 2005). Furthermore, diverse mutant strains, fabricated to evaluate the in vivo function of e-pili, have produced phenotypes consistent with the need for PilA for long-range electron transport. Recently, it has been discovered that the multi heme outer surface c-type cytochrome OmcS of *G. sulfurreducens* can assemble into filaments (Filman et al., 2019; Wang et al., 2019). The impetus for extending research beyond cell-associated cytochromes to extracellular electron transport mechanisms was the discovery that

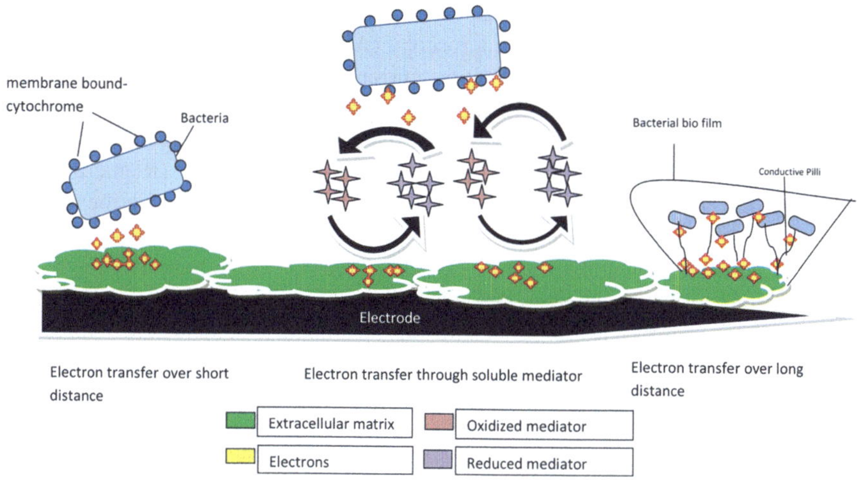


**Fig. 12.1** A pictorial illustration of bacterial electron transport processes

*G. metallireducens* expresses flagella and pili when grown on insoluble Fe(III) or Mn (IV) oxides, but not on soluble Fe(III) citrate (Childers et al., 2002). Transcripts of the putative monomeric pilin gene, PilA, were detected during growth on Fe(III) oxide, but not on soluble Fe(III) (Childers et al., 2002). The same pattern of pili expression was observed during growth on Fe(III) oxide, but not with soluble Fe(III), in *G. sulfurreducens* (Reguera et al., 2005). The diameter of the *G. sulfurreducens* filaments measured in these preliminary studies was 3 nm which was found to be consistent with later measurements of *G. sulfurreducens* e-pili (Malvankar et al., 2014, 2015; Adhikari et al., 2016; Tan et al., 2017). Heterologous expression of the *G. sulfurreducens* PilA pilin monomer in *Pseudomonas aeruginosa* also yields e-pili with a diameter of 3 nm (Liu et al., 2019). The OmcS filament has a diameter of 4 nm (Filman et al., 2019; Wang et al., 2019). The 3 nm diameter of the *G. sulfurreducens* pili is thinner than that of other bacteria, which may be because *G. sulfurreducens* pilin monomer consists of only 61 amino acids, without the large "head group" that is usually found on the outer surface of other type IV pili (Reguera et al., 2005; Malvankar et al., 2015). Although it has been possible to develop strains of *G. sulfurreducens* to produce filaments other than PilA pilin-based pili (Klimes et al., 2010), however the evolved strain was not included in any e-pili studies further. When the pilin monomer gene PilA was deleted from the wild-type strain, no filament was observed (Reguera et al., 2005). It can be simply explained from these observations that the filaments of the wild-type strain of *G. sulfurreducens* are pili comprised of PilA pilin monomers.

Alongwith growth on Fe (III) or Mn (IV) oxides, pili expression could also be initiated in *G. sulfurreducens* by growing it at the suboptimal temperature of 25 °C with fumarate as the electron acceptor (Reguera et al., 2005). On adding crystalline Fe (III) oxide to these fumarate-grown cells, the nano particles preferably attached to the pili. This observation led to the investigation of how cells are able to transfer electrons to Fe (III) oxides when suspended on pili so distantly from the cell surface. Conductive-tip atomic force microscopy (AFM) has shown that the pili were electrically conductive with a linear ohmic current-voltage response (Reguera et al., 2005). These findings, together with the inability of pilA-deficient mutants to reduce Fe (III) oxides, lead to the conclusion that the pili of *G. sulfurreducens* serve as remote "Microbial Nanowires" for extracellular electron transfer (Reguera et al., 2005). Analysis of current generation density on the anode of a bioelectrochemical device revealed that microbial nanowires also play an important role in long-range electron transport through biofilms of *G. sulfurreducens* which generated high current densities (Reguera et al., 2006). Pilin-based microbial nanowires are now referred to as e-pili to avoid confusion with other conductive structures (Lovley, 2017b).

## 12.4 OmcS and OmcS Filaments

Similar to e-pili, cytochrome hexaheme c-type OmcS was unearthed by comparing protein expression in cells cultutred on insoluble Mn(IV) or Fe(III) oxides versus soluble Fe(III) citrate. OmcSare rich in proteins readily sheared from the extracellular surface of cells cultured on Mn(IV) oxide. OmcS gene transcripts were detected by RT-PCR during growth on Fe(III) oxide and fumarate, but not on Fe(III) citrate (Mehta et al., 2005). Like e-pili, deletion of the OmcSgene prevented growth on Mn(IV) and Fe(III) oxides, but growth of soluble extracellular electron acceptors was not affected (Mehta et al., 2005). The outer surface localization of OmcS, together with the earlier observed localization of Fe(III) oxide on pili, led to the idea that OmcSis the transmembrane transporter of c-type cytochromes that transports electrons to the outer surface and the e-pili (Mehta et al., 2005; Lovley, 2006). The possibility that OmcS may form filaments to facilitate long-range electron transport at distance from the cell has been considered in several studies, but ultimately had to be ruled out (Tran, 2009; Leang et al., 2010).

Further, cryo-electron microscopy examination of the filaments on outer surface of *G. sulfurreducens* proteins showed that OmcS can assemble into μm-long filaments (Filman et al., 2019; Wang et al., 2019). Surface proteins are important because one possibility is that the filaments of OmcS may assemble into artifacts generated during harvesting of the outer surface proteins. During purification the highly hydrophobic OmcS and other proteins were isolated from the lipid-rich environment of the extracellular surface and suspended in high-salt and high-pH aqueous solutions (Qian et al., 2011). Changes in chemical conditions are known to lead to the polymerization of c-type cytochrome monomers into nanowires (Hirota et al., 2010; Haldar et al., 2015; Alvarez De Eulate et al., 2017; Hirota, 2019; Nucara et al., 2019). One of the important stabilizing element in the structure of the OmcS filament is coordination of histidine in one subunit with iron in the heme of an adjacent subunit (Filman et al., 2019; Wang et al., 2019). It is observed during the routine purification of proteins that, the his-heme metal arrangement is a very strong bond that can form spontaneously (Bornhorst & Falke, 2000). This coordination between histidine and metals is favored at higher pH, as the outer surface filament preparation methods used to recover OmcS filaments values (Bornhorst & Falke, 2000). OmcS filaments emanating from wild-type *G. sulfurreducens* have not yet been imaged, but evidences indicate expression by a mutant strain that significantly overexpressed OmcS (Wang et al., 2019).

## 12.5 Biological Implications

Microorganisms have reportedly shown the use of nanowires to facilitate the use of extracellular metals as terminal electron acceptors in electron transport chains. The high reduction potential of the metals accepting electrons can lead to a large production of ATP production (Nealson & Rowe, 2016; Shi et al., 2016). Beyond this, the magnitude of the impact caused by the presence by the presence of bacterial nanowires has yet to be fully realized. It has been speculated that nanowires may act as conduits for transport of electrons between different members of the microbial community. This has the potential to allow regulatory feedback or other communication between members of the same microbial species or even different microbial species (Shi et al., 2016; Rabaey & Rozendal, 2010). Some organisms are able to expel and absorb electrons through nanowires (Nealson & Rowe, 2016). Such species are likely be able to oxidize extracellular metals by using them aselectrons or energy sources to facilitate energy consuming cellular processes. Nanowires can be potentially used by microorganisms to temporarily store electrons on metals. The increased concentration of electrons on the metal anode will create a battery that cells can later use to stimulate metabolic activity (Shi et al., 2016). Although these potential effects provide plausible hypotheses about the role of bacterial nanowires in biological systems, further research is needed to fully understand the extent to which cellular species benefit from the use of nanowires (Nealson & Rowe, 2016).

## 12.6 Potential Applications

### *12.6.1 Microbial Fuel Cells*

Microbial nanowires can enable the generation of electricity in microbial fuel cells (MFCs). In an MFC, microbes transfer electrons to an electrode via nanowires, producing electrical current (Fig. 12.2). This can be used in the generation of renewable energy. In microbial fuel cells (MFCs), bacterial nanowires generate electricity bytransporting extracellular electrons to the anode of the MFC (Kodesia et al.,

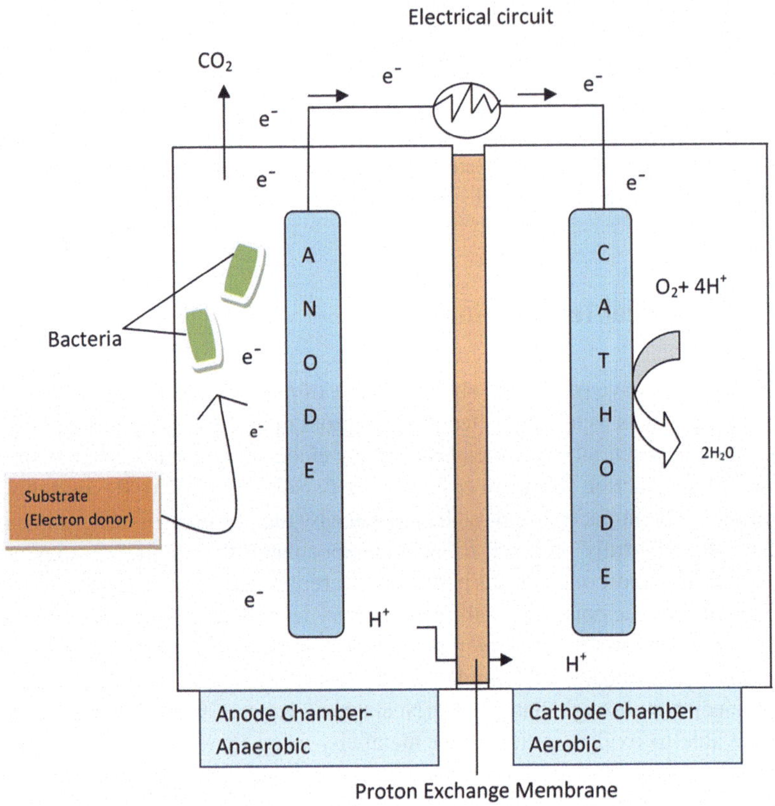


**Fig. 12.2** Microbial fuel cell

2017). Nanowire arrays have been shown to improve the power output of MFCs with efficient and long-range electrical conductivity. In particular, bacterial nanowires from *Geobactersulfurreducens* exhibit metallic conductivity, generating electricity at levels comparable to synthetic metallic nanostructures (Malvankar et al., 2011). Higher energy production has generally been observed when bacterial strains have been genetically modified to promote nanowire formation (Malvankar & Lovley, 2012). Coating nanowires with metal oxides also further promotes electrical conductivity (Maruthupandy et al., 2017). In addition, these nanowires can transport electrons at centimeter-scale distances (Malvankar & Lovley, 2012). Long-range electron transfer through microbial nanowire networks allows living cells that are not in direct contact with the anode to facilitate electron flow (Reguera et al., 2006). The current density has been reported to be about 17 microamps per square centimeter at a voltage of about 0.5 volts on a biofilm 7 microns thick (Liu et al., 2020).

### 12.6.2 Microbial Electrosynthesis

Microbial electrosynthesis (MES) involves the use of microbes to convert carbon dioxide (CO2) and electrical current into useful chemicals, such as methane or other biofuels. In MES, microbes consume electrons from an electrode and use them to drive metabolic reactions, producing the desired product. The overall process is a form of microbial bio-electrosynthesis, as it relies on the transfer of electrons from the electrode to the microbes. Several bioenergetic strategies require microbes to exchange electrons with their external environment. For example, efficient metabolism of alcohol and fatty acids is important for the anaerobic digestion of organic waste into methane, requiring an exchange of electrons between microorganisms that oxidize these substrates and methanogens that accept electrons therby reducing carbon dioxide to methane (Sieber et al., 2012). The conversion of organic waste into electricitythrough microbial fuel cells (MFCs) requires microbes to transfer electrons to electrodes (Lovley, 2012; Logan & Rabaey, 2012). Microbial electrode exchange is also essential for processes such as electromethanogenesis as well as microbial electrosynthesis, where microorganisms reduce carbon dioxide to methane or multicarbon fuels using the electrons obtained from electrodes (Lovley & Nevin, 2013; Rabaey & Rozendal, 2010).

### 12.6.3 Bioremediation

Microbial nanowires can be used to clean up contaminated environments. Certain microbes with nanowires can reduce or oxidize metals and minerals, which can be useful in bioremediation of soil and water. Microbial nanowires from Shewanella and Geobacter contribute to the bioremediation of uranium in underground waters (Jiang et al., 2011). To demonstrate this, scientists compared and observed the concentrations of uranium eliminated by hairy and glabrous strains of Geobacter. Through a series of controlled experiments, they were able to conclude that the strain in which the nanowires were present was more efficient for uranium mineralization compared to the mutant without the nanowires (Cologgi et al., 2011). The natural ability of Geobacter pili to bind and reduce deposited cationic metal contaminants such as uranyl cations is of particular interest for bioremediation applications (Cologgi et al., 2011). Sensors can be developed with ability to exploit the planar self-assembly of pilin derivatives, their electrical conductivity, and the concentration of surface metal traps to increase detection sensitivity (Cosert et al., 2017). Since metal binding is mediated by acidic amino acid ligands, genetic engineering can be used to fine-tune binding kinetics and specificity and achieve the selectivity required for environmental sensing applications. Such a platform could enable the bioremediation of environments affected by a variety of cationic metals, including uranium, cadmium and cobalt.

### 12.6.4 Bioelectronics

Another important application of bacterial nanowires can be seen in the bioelectronics industry (Sure et al., 2016). Microbial nanowires can be used as biological components in electronic devices. They can be integrated with electronic circuits to create bioelectronic devices that can perform various functions such as sensing, computing, and data storage. Considering the sustainable resources, scientists have proposed the future use of Geobacter biofilms as a platform for functional underwater transistors and supercapacitors capable of self-renewable energy sources (Malvankar & Lovley, 2012). Researchers have highlighted a diffusive memristor made up of protein nanowires of *Geobacter sulfurreducens* which functions at sufficiently lower voltages than the ones described earlier. This may allow the construction of artificial neurons which function usually at voltages of biological action potentials. Bacterial nanowires differ from traditional nanowires by showing a higher degree of biocompatibility with silicon nanowires. Further research is needed, but eventually memristors could be used to directly process biosensing signals for computational neuromorphic applications and/or direct communication with biological neurons (Fu et al., 2020).

### 12.6.5 Biosensors

Nanowires have substantial potential as the sensor component in electronic sensing devices. However, surface functionalization of traditional nanowire and nanotube materials with short peptides that increase sensor selectivity and sensitivity requires complex chemistries with toxic reagents. In contrast, microorganisms can assemble pilin monomers into protein nanowires with intrinsic conductivity from renewable feedstocks, yielding an electronic material that is robust and stable in applications, but also biodegradable (Lekbach et al., 2023). Microbial nanowires can be used to create biosensors. For example, by attaching specific enzymes or antibodies to the nanowires, they can detect and bind to specific molecules, enabling the detection of pathogens or environmental pollutants.

### 12.6.6 Biotechnology

Microbial nanowires can be used in biotechnology for the production of high-value products such as biofuels, chemicals, and pharmaceuticals (Fig. 12.3). By engineering microbes with specific properties, they can be used to synthesize valuable compounds using nanowires to transfer electrons. Electroactive bacteria

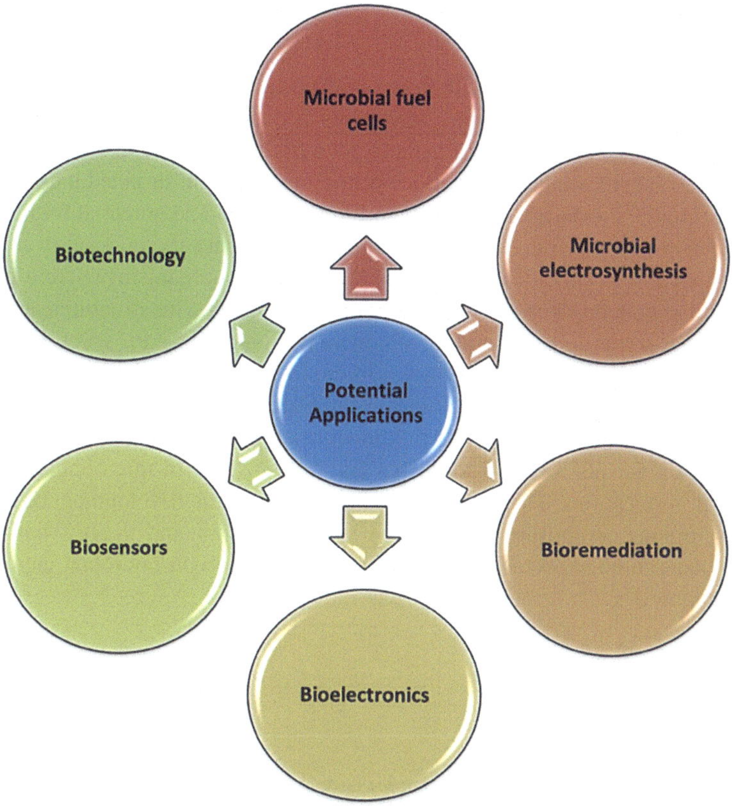


**Fig. 12.3** Potential applications of microbial nanowires

can perform bidirectional extracellular electron transfer (EET), exchanging electrons and energy with the extracellular environment, thus playing a central role in the process of microbial electrofermentation (EF). Imbalance fermentation and microbial electrosynthesis are major pathways for the production of valueadded chemicals and biofuels. EF is a new fermentation process, which can use electron transfer to promote substance metabolism by changing intracellular redox potential (ORP) and NADH/NAD+ ratio, etc. (Moscoviz et al., 2016). However, the productivity and efficiency achieved by the above systems are still too low for industrial-scale production (Schievano et al., 2016). The main bottlenecks of EF are the insufficient rate of EET, the lack of gene editing tools available to design the metabolic pathways of the target products and the manipulation of electrode materials and bioelectrochemical reactors, all of which limit the large-scale application of electrofermentation.

## 12.7 Conclusion

In addition to their microbial role, microbial protein nanowires with biologically produced wires or their biomimetic derivatives as conductive elements in electronic devices have attracted attention for their potential. Various studies on e-pili have shown that the conductance and surface properties of single stranded nanowire can be tuned through the design of synthetic pilin monomers. The potential for heterologous expressionof e-pili in other microorganisms provides the possibility of large-scale production in specifically designed scaffold microorganismswhere e-pili are the only filaments produced. OmcS filaments offer a nanowire structures and properties distinct from e-pili, expanding nanowire design options. Prospecting for new types of nanowires in the microbial world could expand the diversity of material properties available for electronic innovation. Further elucidation of the structures and electron-transport mechanisms in known microbial nanowires is likely to lead to advanced concepts for rational design and optimization of microbial nanowires for further applications. There are exciting potentials in the development of durable and sustainable composite materials, wearable sensors and energy harvesting devices that can be expected to promote further research on microbial engineered protein nanowires.

## References

Adhikari, R. Y., Malvankar, N. S., Tuominen, M. T., & Lovley, D. R. (2016). Conductivity of individual Geobacter pili. *RSC Advances, 6*, 8354–8357. https://doi.org/10.1039/C5RA28092C

Alvarez De Eulate, E., O'sullivan, S., & Arrigan, D. W. M. (2017). Electrochemically induced formation of cytochrome c oligomers at soft interfaces. *ChemElectroChem, 4*, 898–904. https://doi.org/10.1002/celc.201600851

Bornhorst, J. A., & Falke, J. J. (2000). Purification of proteins using polyhistidine affinity tags. *Methods in Enzymology, 326*, 245–254. https://doi.org/10.1016/S0076-6879(00)26058-8

Brutinel, E., & Gralnick, J. (2012). Shuttling happens: Soluble flavin mediators of extracellular electron transfer in Shewanella. *Applied Microbiology and Biotechnology, 93*, 41–48.

Childers, S. E., Ciufo, S., & Lovley, D. R. (2002). *Geobactermetallireducens* accesses insoluble Fe(III) oxide by chemotaxis. *Nature, 416*, 767–769. https://doi.org/10.1038/416767a

Cologgi, D. L., Lampa-Pastirk, S., Speers, A. M., Kelly, S. D., & Reguera, G. (2011). Extracellular reduction of uranium via Geobacter conductive pili as a protective cellular mechanism. *Proceedings of the National Academy of Sciences of the United States of America, 108*(37), 15248–15252.

Cosert, K. M., Steidl, R. J., Castro-Forero, A., Worden, R. M., & Reguera, G. (2017). Electronic characterization of *Geobactersulfurreducens* pilins in self-assembled monolayers unmasks tunnelling and hopping conduction pathways. *Physical Chemistry Chemical Physics, 19*, 11163–11172.

Creasey, R. C. G., Mostert, A. B., Nguyen, T. A. H., Virdis, B., Freguia, S., & Laycock, B. (2018). Microbial nanowires–electron transport and the role of synthetic analogues. *Acta Biomaterialia, 69*, 1–30. https://doi.org/10.1016/j.actbio.2018.01.007

Filman, D. J., Marino, S. F., Ward, J. E., Yang, L., Mester, Z., Bullitt, E., et al. (2019). Cryo-EM reveals the structural basis of long-range electron transport in a cytochrome-based bacterial nanowire. *Communications Biology, 2*, 219. https://doi.org/10.1038/s42003-019-0448-9

Fu, T., Liu, X., Gao, H., Ward, J. E., Liu, X., Yin, B., et al. (2020). Bioinspired bio-voltage memristors. *Nature Communications, 11*(1), 1861. https://doi.org/10.1038/s41467-020-15759-y. Bibcode:2020NatCo..11.1861F. PMC 7171104. PMID 32313096.

Gorby, Y., & Lovley, D. R. (1991). Electron transport in the dissimilatory iron-reducer, GS-15. *Applied and Environmental Microbiology, 57*, 867–870.

Gorby, Y. A., Yanina, S., McLean, J. S., Rosso, K. M., Moyles, D., Dohnalkova, A., Beveridge, T. J., Chang, I. S., Kim, B. H., & Kim, K. S. (2006). Electrically conductive bacterial nanowires produced by Shewanellaoneidensis strain MR-1 and other microorganisms. *Proceedings of the National Academy of Sciences of the United States of America, 103*, 11358–11363.

Gutermann, T., & Gazit, E. (2018). Toward peptide-based bioelectronics: reductionist design of conductive pili mimetics. *Bioelectron Med, 1*, 131–137. https://doi.org/10.2217/bem-2018-0003

Haldar, S., Sil, P., Thangamuniyandi, M., & Chattopadhyay, K. (2015). Conversion of amyloid fibrils of cytochrome c to mature nanorods through a honeycomb morphology. *Langmuir, 31*, 4213–4223. https://doi.org/10.1021/la5029993

Hirota, S. (2019). Oligomerization of cytochrome c, myoglobin, and related heme proteins by 3D domain swapping. *Journal of Inorganic Biochemistry, 194*, 170–179. https://doi.org/10.1016/j.jinorgbio.2019.03.002

Hirota, S., Hattori, Y., Nagao, S., Taketa, M., Komori, H., Kamikubo, H., et al. (2010). Cytochrome c polymerization by successive domain swapping at the C-terminal helix. *Proceedings of the National Academy of Sciences of the United States of America, 107*, 12854–12859. https://doi.org/10.1073/pnas.1001839107

Holmes, D. E., Dang, Y., Walker, D. J. F., & Lovley, D. R. (2016). The electrically conductive pili of Geobacter species are a recently evolved feature for extracellular electron transfer. *Microbial Genomics, 2*, e000072. https://doi.org/10.1099/mgen.0.000072

Ing, N. L., El-Naggar, M. Y., & Hochbaum, A. I. (2018). Going the distance: Long-range conductivity in protein and peptide bioelectronic materials. *The Journal of Physical Chemistry. B, 122*, 1043–10423. https://doi.org/10.1021/acs.jpcb.8b07431

Inoue, K., Leang, C., Franks, A. E., Woodard, T. L., Nevin, K. P., & Lovley, D. R. (2010). Specific localization of the c-type cytochrome OmcZ at the anode surface in current-producing biofilms of Geobactersulfurreducens. *Environmental Microbiology Reports, 3*, 211–217.

IPCC. (2014). Climate Change 2014: Mitigation of Climate Change. In O. Edenhofer, R. Pichs-Madruga, Y. Sokona, E. Farahani, S. Kadner, K. Seyboth, A. Adler, I. Baum, S. Brunner, P. Eickemeier, B. Kriemann, J. Savolainen, S. Schlömer, C. von Stechow, T. Zwickel, & J. C. Minx (Eds.), *Contribution of Working Group III to the Fifth Assessment Report of the Intergovernmental Panel on Climate Change* (pp. 7–9). Cambridge University Press.

Jiang, S., Kim, M. G., Kim, S. J., Jung, H. S., Lee, S. W., Noh, D. Y., et al. (2011). Bacterial formation of extracellular U(VI) nanowires. *Chemical Communications, 47*(28), 8076–8078. https://doi.org/10.1039/C1CC12554K

Kato, S., Hashimoto, K., & Watanabe, K. (2012a). Microbial interspecies electron transfer via electric currents through conductive minerals. *Proceedings of the National Academy of Sciences of the United States of America, 109*, 10042–10046.

Kato, S., Hashimoto, K., & Watanabe, K. (2012b). Methanogenesis facilitated by electric syntrophy via (semi)conductive iron-oxide minerals. *Environmental Microbiology, 14*, 1646–1654.

Klimes, A., Franks, A. E., Glaven, R. H., Tran, H. T., Barrett, C. L., Qiu, Y., et al. (2010). Production of pilus-like filaments in *Geobacter sulfurreducens* in the absence of the type IV PilA protein PilA. *FEMS Microbiology Letters, 310*, 62–68. https://doi.org/10.1111/j.1574-6968.2010.02046.x

Kodesia, A., Ghosh, M., & Chatterjee, A. (2017). *Development of biofilm nanowires and electrode for efficient Microbial Fuel Cells (MFCs)*. Thapar University Digital Repository (TuDR).

Kotloski, N. J., & Gralnick, J. A. (2013). Flavin electron shuttles dominate extracellular electron transfer by Shewanellaoneidensis. *MBio, 4*, e00553–e00512.

Kracke, F., Vassilev, I., & Krömer, J. O. (2015). Microbial electron transport and energy conservation–the foundation for optimizing bioelectrochemical systems. *Frontiers in Microbiology, 6*, 575. https://doi.org/10.3389/fmicb.2015.00575

Leang, C., Qian, X., Mester, T., & Lovley, D. R. (2010). Alignment of the c-type cytochrome OmcS along pili of Geobactersulfurreducens. *Applied and Environmental Microbiology, 76*, 4080–4084.

Lekbach, Y., Ueki, T., Liu, X., Woodard, T., Yao, J., & Lovley, D. R. (2023). Microbial nanowires with genetically modified peptide ligands to sustainably fabricate electronic sensing devices. *Biosensors and Bioelectronics, 226*, 115147. https://doi.org/10.1016/j.bios.2023.115147

Liao, J. C., Mi, L., Pontrelli, S., & Luo, S. (2016). Fuelling the future: Microbial engineering for the production of sustainable biofuels. *Nature Reviews. Microbiology, 14*, 288–304. https://doi.org/10.1038/nrmicro.2016.32I

Liu, F., Rotaru, A.-E., Shrestha, P. M., Malvankar, N. S., Nevin, K. P., & Lovley, D. R. (2012). Promoting direct interspecies electron transfer with activated carbon. *Energy & Environmental Science, 5*, 8982–8989.

Liu, X., Wang, S., Xu, A., Zhang, L., Liu, H., & Ma, L. Z. (2019). Biological synthesis of high-conductive pili in aerobic bacterium *Pseudomonas aeruginosa. Applied Microbiology and Biotechnology, 103*, 1535–1544. https://doi.org/10.1007/s00253-018-9484-5

Liu, X., Gao, H., Ward, J. E., Liu, X., Yin, B., Fu, T., et al. (2020). Power generation from ambient humidity using protein nanowires. *Nature, 578*(7796), 550–554. Bibcode:2020Natur.578..550L. https://doi.org/10.1038/s41586-020-2010-9

Logan, B. E., & Rabaey, K. (2012). Conversion of wastes into bioelectricity and chemicals by using microbial electrochemical technologies. *Science, 337*, 686–690.

Lovley, D. R. (2006). Bug juice: Harvesting electricity with microorganisms. *Nature Reviews Microbiology, 4*, 497–508. https://doi.org/10.1038/nrmicro1442

Lovley, D. R. (2012). Electromicrobiology. *Annual Review of Microbiology, 66*, 391–409.

Lovley, D. R. (2017a). E-biologics: Fabrication of sustainable electronics with 'green' biological materials. *mBio, 8*, e00695–e00617. https://doi.org/10.1128/mBio.00695-17

Lovley, D. R. (2017b). Electrically conductive pili: Biological function and potential applications in electronics. *Current Opinion in Electrochemistry, 4*, 190–198. https://doi.org/10.1016/j.coelec.2017.08.015

Lovley, D. R., & Nevin, K. P. (2013). Electrobiocommodities: Powering microbial production of fuels and commodity chemicals from carbon dioxide with electricity. *Current Opinion in Biotechnology, 24*, 385–390.

Lovley, D. R., Giovannoni, S. J., White, D. C., Champine, J. E., Phillips, E. J. P., Gorby, Y. A., et al. (1993). *Geobactermetallireducens* gen. nov. sp. nov., a microorganism capable of coupling the complete oxidation of organic compounds to the reduction of iron and other metals. *Archives of Microbiology, 159*, 336–344. https://doi.org/10.1007/BF00290916

Lovley, D. R., Holmes, D. E., & Nevin, K. P. (2004). Dissimilatory Fe(III) and Mn(IV) reduction. *Advanced in Microbial Physiology, 49*, 219–286. https://doi.org/10.1016/S0065-2911(04)49005-5

Malvankar, N. S., & Lovley, D. R. (2012). Microbial nanowires: A new paradigm for biological electron transfer and bioelectronics. *ChemSusChem, 5*(6), 1039–1046. https://doi.org/10.1002/cssc.201100733. PMID 22614997

Malvankar, N. S., Vargas, M., Nevin, K. P., Franks, A. E., Leang, C., Kim, B. C., Inoue, K., Mester, T., Covalla, S. F., Johnson, J. P., Rotello, V. M., Tuominen, M. T., & Lovley, D. R. (2011). Tunable metallic-like conductivity in microbial nanowire networks. *Nature Nanotechnology, 6*(9), 573–579. Bibcode:2011NatNa...6..573M. https://doi.org/10.1038/nnano.2011.119

Malvankar, N. S., Yalcin, S. E., Tuominen, M. T., & Lovley, D. R. (2014). Visualization of charge propagation along individual pili proteins using ambient electrostatic force microscopy. *Nature Nanotechnology, 9*, 1012–1017. https://doi.org/10.1038/nnano.2014.236

Malvankar, N. S., Vargas, M., Nevin, K. P., Tremblay, P.-L., Evans-Lutterodt, K., Nykypanchuk, D., et al. (2015). Structural basis for metallic-like conductivity in microbial nanowires. *mBio, 6*, e00084–e00015. https://doi.org/10.1128/mBio.00084-15

Maruthupandy, M., Anand, M., Maduraiveeran, G., Beevi, A. S., & Priya, R. J. (2017). Fabrication of CuO nanoparticles coated bacterial nanowire film for a high-performance electrochemical conductivity. *Journal of Materials Science, 52*(18), 10766–10778. https://doi.org/10.1007/s10853-017-1248-6. Bibcode:2017JMatS..lo5210766M. S2CID 103105219

Mehta, T., Coppi, M. V., Childers, S. E., & Lovley, D. R. (2005). Outer membrane c-type cytochromes required for Fe(III) and Mn(IV) oxide reduction in *Geobactersulfurreducens. Applied and Environmental Microbiology, 71*, 8634–8641. https://doi.org/10.1128/AEM.71.12.8634-8641.2005

Moscoviz, R., Toledo-Alarcón, J., Trably, E., & Bernet, N. (2016). Electro-fermentation: How to drive fermentation using electrochemical systems. *Trends in Biotechnology, 34*(11), 856–865. https://doi.org/10.1016/j.tibtech.2016.04.009

Nealson, K. H., & Rowe, A. R. (2016). Electromicrobiology: Realities, grand challenges, goals and predictions. *Microbial Biotechnology, 9*(5), 595–600. https://doi.org/10.1111/1751-7915.12400

Nucara, A., Carbone, M., Ripanti, F., Manganiello, R., Postorino, P., & Caronaro, M. (2019). Achieving cytochrome c fibril/aggregate control towards micro-platelets and micro-fibers by tuning pH and protein concentration: A combined morphological and spectroscopic analysis. *International Journal of Biological Macromolecules, 138*, 106–115. https://doi.org/10.1016/j.ijbiomac.2019.07.060

Orellana, R., Leavitt, J. J., Comolli, L. R., Csencsits, R., Janot, N., Flanagan, K. A., Gray, A. S., Leang, C., Izallalen, M., & Mester, T. (2013). U(VI) reduction by a diversity of outer surface c-type cytochromes of Geobactersulfurreducens. *Applied and Environmental Microbiology, 79*(20), 6369–6374.

Qian, X., Mester, T., Morgado, L., Arakawa, T., Sharma, M. L., Inoue, K., et al. (2011). Biochemical characterization of purified OmcS, a C-type cytochrome required for insoluble Fe(III) reduction in *Geobactersulfurreducens. Biochimica et Biophysica Acta, 1807*, 404–412. https://doi.org/10.1016/j.bbabio.2011.01.003

Rabaey, K., & Rozendal, R. A. (2010). Microbial electrosynthesis – Revisiting the electrical route for microbial production. *Nature Reviews Microbiology, 8*(10), 706–716.

Reguera, G., McCarthy, K. D., Mehta, T., Nicoll, J. S., Tuominen, M. T., & Lovley, D. R. (2005). Extracellular electron transfer via microbial nanowires. *Nature, 435*, 1098–1101.

Reguera, G., Nevin, K. P., Nicoll, J. S., Covalla, S. F., Woodard, T. L., & Lovley, D. R. (2006). Biofilm and nanowire production leads to increased current in *Geobactersulfurreducens* fuel cells. *Applied and Environmental Microbiology, 72*, 7345–7348. https://doi.org/10.1128/AEM.01444-06

Schievano, A., PepéSciarria, T., Vanbroekhoven, K., De Wever, H., Puig, S., Andersen, S. J., Rabaey, K., & Pant, D. (2016). Electro-fermentation – Merging electrochemistry with fermentation in industrial applications. *Trends in Biotechnology, 34*(11), 866–878. https://doi.org/10.1016/j.tibtech.2016.04.007

Shi, L., Dong, H., Reguera, G., Beyenal, H., Lu, A., Liu, J., et al. (2016). Extracellular electron transfer mechanisms between microorganisms and minerals. *Nature Reviews. Microbiology, 14*, 651–662. https://doi.org/10.1038/nrmicro.2016.93

Sieber, J. R., McInerney, M. J., & Gunsalus, R. P. (2012). Genomic insights into syntrophy: The paradigm for anaerobic metabolic cooperation. *Annual Review of Microbiology, 66*, 429–452.

Sure, S., Ackland, M. L., Torriero, A. J., Adholeya, A., & Kochar, M. (2016). Microbial nanowires: An electrifying tale. *Microbiology, 162*, 2017–2028. https://doi.org/10.1099/mic.0.000382

Tan, Y., Adhikari, R. Y., Malvankar, N. S., Ward, J. E., Woodard, T. L., Nevin, K. P., et al. (2017). Expressing the *Geobactermetallireducens* PilA in *Geobactersulfurreducens* yields pili with exceptional conductivity. *mBio, 8*, e02203–e02216. https://doi.org/10.1128/mBio.02203-16

Tran, H. T. (2009). Investigation of chemotaxis genes and their functions in Geobacter species. *Open Access Dissertations, 129*. Available at: https://scholarworks.umass.edu/open_access_dissertations/129

Ueki, T., Walker, D. J. F., Tremblay, P.-L., Nevin, K. P., Ward, J. E., Woodard, T. L., et al. (2019). Decorating the outer surface of microbially produced protein nanowires with peptides. *ACS Synthetic Biology, 8*, 1809–1817. https://doi.org/10.1021/acssynbio.9b00131

Voordeckers, J. W., Kim, B.-C., Izallalen, M., & Lovley, D. R. (2010). Role of geobactersulfurreducens outer surface c-type cytochromes in reduction of soil humic acid and anthraquinone-2,6- disulfonate. *Applied and Environmental Microbiology, 76*, 2371–2375.

Walker, D. J. F., Adhikari, R. Y., Holmes, D. E., Ward, J. E., Woodard, T. L., Nevin, K. P., et al. (2018). Electrically conductive pili from genes of phylogenetically diverse microorganisms. *The ISME Journal, 12*, 48–58. https://doi.org/10.1038/ismej.2017.141

Walker, D. J. F., Martz, E., Holmes, D. E., Zhou, Z., Nonnenmann, S. S., & Lovley, D. R. (2019). The archaellum of *Methanospirillumhungatei* is electrically conductive. *mBio, 10*, e00579–e00519. https://doi.org/10.1128/mBio.00579-19

Wang, F., Gu, Y., O'brien, J. P., Yi, S. M., Yalcin, S. E., Srikanth, V., et al. (2019). Structure of microbial nanowires reveals stacked hemes that transport electrons over micrometers. *Cell, 177*, 361–369. https://doi.org/10.1016/j.cell.2019.03.029

# Index




P. Saini, P. Mishra, *Bioprospecting of Microbial Resources for Agriculture, Environment and Bio-chemical Industry*,
https://doi.org/10.1007/978-3-031-63844-2

# GPSR Compliance

*The European Union's (EU) General Product Safety Regulation (GPSR) is a set of rules that requires consumer products to be safe and our obligations to ensure this.*

*If you have any concerns about our products, you can contact us on ProductSafety@springernature.com*

In case Publisher is established outside the EU, the EU authorized representative is:

Springer Nature Customer Service Center GmbH
Europaplatz 3
69115 Heidelberg, Germany

The manufacturer's authorised representative in the EU is Springer Nature Customer Service Centre GmbH, Europaplatz 3, 69115 Heidelberg, Germany. If you have any concerns regarding our products, please contact ProductSafety@springernature.com

Printed and bound by CPI Group (UK) Ltd, Croydon, CR0 4YY
25/09/2026
02222892-0004